Heritage Masonry

Materials and Structures

Heritage Masonry

Materials and Structures

Edited by

S. Syngellakis
Wessex Institute of Technology, UK

WIT*PRESS* **Southampton, Boston**

Editor:

S. Syngellakis
Wessex Institute of Technology, UK

Published by

WIT Press
Ashurst Lodge, Ashurst, Southampton, SO40 7AA, UK
Tel: 44 (0) 238 029 3223; Fax: 44 (0) 238 029 2853
E-Mail: witpress@witpress.com
http://www.witpress.com

For USA, Canada and Mexico

WIT Press
25 Bridge Street, Billerica, MA 01821, USA
Tel: 978 667 5841; Fax: 978 667 7582
E-Mail: infousa@witpress.com
http://www.witpress.com

British Library Cataloguing-in-Publication Data

A Catalogue record for this book is available
from the British Library

ISBN: 978-1-84564-839-8
eISBN: 978-1-84564-840-4

Library of Congress Catalog Card Number: 2013943639

Printed by Lightning Source, UK.

Preface

Masonry is a traditional, highly durable mode of construction; many heritage masonry structures, built at various historical periods, have survived, to a less or greater extent, adverse environmental conditions, which have reduced, sometimes considerably, their integrity, strength and durability. Due to the cultural significance of heritage architecture, resources are today allocated towards their restoration and conservation. This volume comprises distinguished contributions from the Transactions of the Wessex Institute describing research efforts towards achieving these objectives.

Knowledge and understanding of constituent materials, modes of construction and overall mechanical behaviour precedes any restorative action. This can be attained through architectural surveys, recourse to documented evidence, laboratory investigations and in situ tests. In relation to dynamic behaviour, which can be critical to masonry built in seismic areas, sonic pulse velocity tests and micro-vibration measurements are particularly relevant. A collection of significant amount of data on possible failure mechanisms can lead to a reliable probabilistic model for structural strength assessment.

Research also focuses on the binding material mixtures. Laboratory tests provide information on the composition and properties of ancient mortars; such results inform the design of contemporary repair material compatible with the original building blocks. In situ infra-red thermography measurements describe the condition and substrate of existing mortar coatings.

As an alternative to the in situ, non-destructive evaluation of existing structures, purpose-built wall specimens are tested in the laboratory. Apart of the material characterisation of wall assemblies and their constituents, the geometric arrangement of the latter is also looked into since it affects the quality of restored masonry. The experimental investigations extend to the effectiveness and suitability of various contemporary binding materials, such as mortars, plasters, renders and grouts, used in restorations and maintenance of heritage masonry; their composition details can be documented and linked to their performance as restoration materials.

Modelling contributes to preservation tasks by indicating possible modes of structural behaviour and consequent failure mechanisms. Experimental results enhance the reliability of the input to numerical analyses. Aggregate masonry properties can be derived from those of individual constituents through homogenisation. Model calibration and validation is attained by comparison of analytical predictions with experimental measurements, obtained either in situ or in the laboratory.

Analytical investigations are mainly concerned with dynamic behaviour since masonry is more likely to sustain damage under seismic conditions. Linear elastic structural models lead to free vibration modes and frequencies while nonlinear elasto-plastic modelling provides time histories of responses that may lead to failure. The time dependence of masonry under quasi-static conditions is confirmed experimentally and explored analytically through a viscoelastic model.

The various issues mentioned above are addressed by the present collection of scientific papers with considerable insight and thoroughness. It is thus hoped that this volume will fill a gap in the literature as a valuable source of information and guidance to researchers and engineers working in the area of restoration and conservation of heritage masonry structures.

Stavros Syngellakis
The New Forest, 2013

Contents

Methodologies for the evaluation of seismic vulnerability of complex masonry buildings: case histories in the historic centre of Sulmona

L. Binda, A. Anzani & G. Cardani
D.I.S – Politecnico di Milano, Italy

Abstract

The paper describes an investigation, based on detailed knowledge of the material, on the morphology and construction aspects of masonry structures, the mechanical behaviour and the possible failure mechanisms of complex buildings in Sulmona (AQ), an important historic centre in Central Italy. It is an area of great interest, not only for its seismic history, but also because of the characteristics of its prestigious buildings, mainly made of multiple leaf stone masonry. In order to understand the behaviour of the historical buildings, use has been made of a direct survey of the constructions (based on a geometrical survey, a survey of the materials and of their state of damage), indirect information deduced from documentary sources, and data collected through in-situ non destructive or minor destructive testing and laboratory investigation for a chemical, physical and mechanical characterisation of the masonry and its components. Record templates and forms, some of them purposely set up, have been adopted and collected into a database, for damage and masonry quality assessment and evaluation of the future vulnerability of the buildings to in-plane and out-of-plane actions according to damage tables previously produced.

Keywords: ND evaluation, multiple leaf masonry, cross section morphology, historical complex buildings.

1 Introduction

The analysis suggested by the present Italian Seismic Code (OPCM 3274/2003 [1] and subsequent modifications) for the assessment of existing buildings is considered excessively laborious for the intervention on a single building, especially if it rises within a complex, a very common condition in historical centres. A compromise is therefore necessary between what it is reasonably required from a designer and what is essential from an exhaustive assessment point of view. The problem can be addressed by a method that can be easily applied through record templates and forms and allows the strength and vulnerability characteristics of a building to be individualised, which is crucial towards the assessment of structural seismic response.

A multi-level procedure has been set up and applied here to study complex buildings in the historical centre of Sulmona (AQ). The investigation method adopted by the authors since the beginning of the '90s [2] is based on the principle that knowledge is fundamental for the choice of suitable techniques and materials aimed at the preservation and damage prevention of cultural heritage.

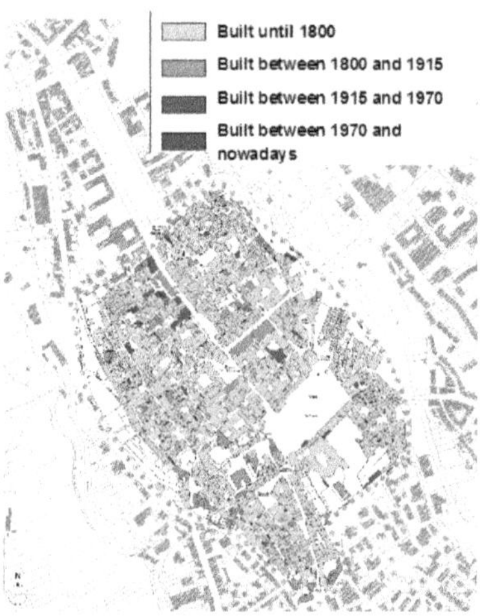

Figure 1: Historic evolution of the historic centre of Sulmona.

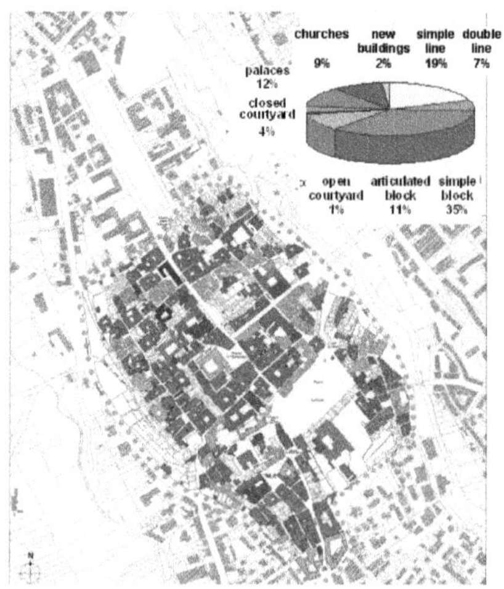

Figure 2: Distribution of building typologies in the centre of Sulmona.

This procedure has been well calibrated through various on site investigations on historic masonry buildings in the Umbria region after the earthquake of 1997 [3], in the Liguria region hit by earthquakes in the 19th century, in the area of Garda Lake hit by the earthquake of 2004 [4, 5]. The earthquake that struck Umbria and Marche regions in 1997 provided the awareness that the lack of knowledge in the material and structural behaviour of the existing buildings was, and still is, the main cause of inappropriate choices of intervention techniques. The problem of repair and retrofitting should be approached in a multidisciplinary way, considering different complementary aspects including: historical evolution of the buildings, geometry and crack pattern, material characteristics, technology of construction and possible failure mechanisms [6]. In this respect, the collaboration between architects/"restorators", historians, structural engineers is particularly important. All collected complementary data will allow an interpretation of the structural response which takes account of the examined masonry and building typology, leading to the evaluation of its seismic vulnerability and the identification of reliable retrofitting procedures for stone-masonry buildings.

2 Study of the historic centre of Sulmona

The first phase of the research presented here consisted of a study of the historical centre of Sulmona aimed at understanding the rules of the formation and growth of the built types, starting from the first construction until the fusion and complex additions of new bodies. The city was founded by the Romans and there was subsequent development in Medieval times. In fig. 1, its more recent evolution is shown where it appears that the main part of the development was carried out mostly before 1800. After collecting historical information on the evolution of the ancient city centre, the elements having structural relevance were analysed.

Diffuse architecture, often poorly constructed, is mainly composed of stonework buildings with timber roofs and floors. In spite of the use of apparently similar material and construction techniques, building characteristics vary according to their typology – from isolated to arrayed buildings – and to their location – from flat to steep mountain sites. This variety is directly connected to the orographical profile of the site and also to the common seismic history.

The recognition of the building typologies and of the characteristic constructive elements is a fundamental aspect aimed at their typo-morphological classification and the creation of a catalogue of the constructive elements characteristic of the local architecture.

The importance of recognizing different building typologies, in view of the definition of their seismic vulnerability, depends on the correspondence between construction typologies and mechanisms of damage. Considering complex cases regarding geometry and original destination, it has to be pointed out that any typology, such as palaces, religious buildings, towers, castles, fortifications, and churches, displays specific problems, also connected to the function of the building itself. Such differences, that have influenced the original construction solutions and the wall masonry quality, similarly will influence the safety assessment and the techniques of intervention. Therefore, for any typology, the key should be to individualise and manage specific levels of investigation, modelling, verification and specific repair techniques, satisfying also the function requirements.

In fig. 2 the location of buildings erected in the 20th century, together with that of historical building typologies have been mapped, consisting of simple and double lines, simple and linked blocks, open and closed courtyards, palaces and churches. It appears that the area of the city centre is mainly occupied by linked blocks and also that many churches are present. Most of the buildings designated "palaces" resulted from the evolution and the fusion of different residential units initially conformed as simple or double lines, mainly during the XVI century, improved through inner courtyards and decorative fixtures.

Subsequently, a template aimed at "Masonry quality evaluation" was adopted and applied to selected buildings characterized by facades with no plaster finishes, therefore having their stonework masonry texture available to visual inspection (figs. 3 and 4). The prediction of the seismic vulnerability of stonework masonry, and particularly of the multiple wythe one, can only be performed provided the intrinsic characteristics of the masonry have been accurately detected. The behavior of masonry highly depends on the construction technique and this is evident from the section layout. In fact, the outer face frequently does not reveal how the masonry section is built since apparently similar facing textures (e.g. regular bonding) may correspond to different types of sections, as one or multiple wythes [7]. A systematic study of the mechanical behaviour of stonework masonry begins from an extensive investigation taking into account the different layers constituting the wall and the kind of constraints which may or may not exist between the layers themselves. From the analysis of this information, constitutive laws for modeling the masonry behavior and for designing possible repair interventions (e.g. injection of grout) can be selected. Input data for the structural analysis with numerical methods are the section survey and the type of connection between the layers. These features can also be taken into account when modeling the in-plane or out of plane failure mechanisms.

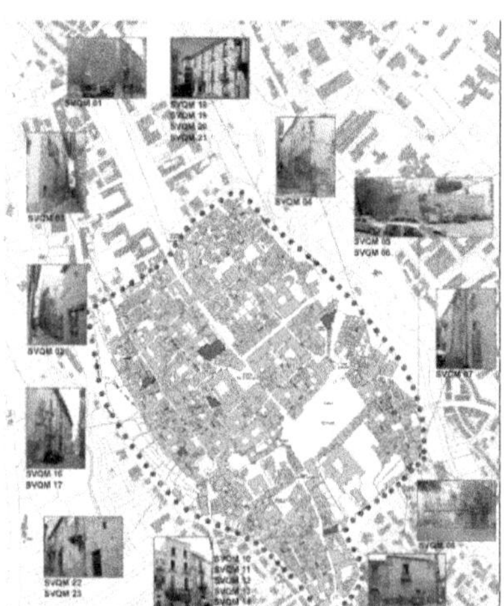

Figure 3: Recognition of building facades without plaster for masonry qualification.

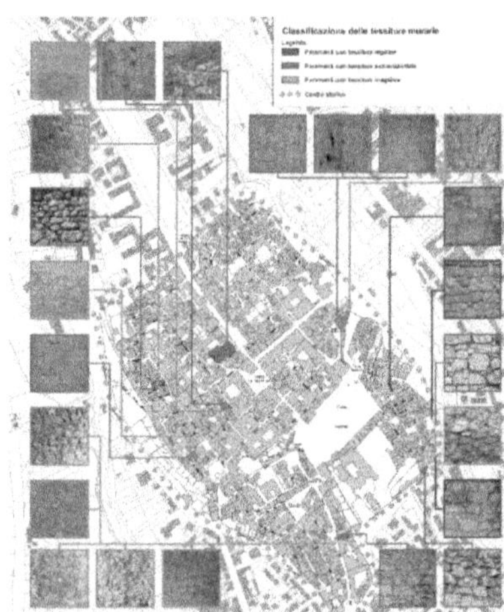

Figure 4: Studied masonry typologies.

Most of the historical buildings in Sulmona turn out to be characterized by three leaf stonework masonry (fig. 5). Considering the masonry textures on the wall facades, they mainly consist of disordered rubble stone rubble (according to table 11.D.1 of the Italian Seismic Code 3431/2005 [8]), always of calcareous nature, with friable mortar. The observed textures are generally made of irregular and sub-horizontal courses, in few cases of regular ones, and with unfinished or partially finished stone blocks. The stone interlocking on the outer face is generally low or very low, in few cases medium. In general the masonry quality, evaluated by visual inspection, turns out to be scarce. On some of these buildings, also a second template aimed to the "Typological and seismic damage survey of buildings" and the evaluation of their vulnerability was applied. Both templates had been set up within the framework of the Italian Network of University Laboratory of Seismic Engineering (ReLUIS).

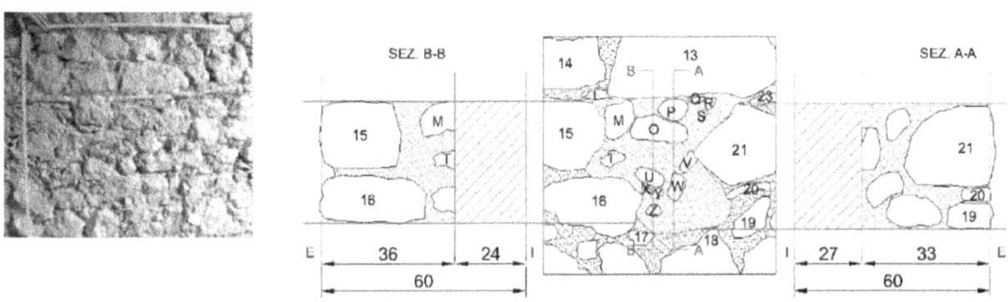

Figure 5: Three-leaf masonry section on block 39 in Ovidio street.

3 Direct investigation on selected case studies

Significant building complexes were identified (painted red in fig. 3 and blue in fig. 4) and subjected to a systematic investigation which included the following phases: (i) geometrical survey of the building and of its crack pattern, (ii) interpretation of the crack pattern and definition of the damage or collapse mechanisms affecting each building, (iii) detection of the connections between walls and between roofs and vaults to walls, (iv) non destructive evaluation of damage, (v) survey of the masonry texture and of the morphology of the wall sections, (vi) on-site characterisation of the masonry walls through sonic and flat-jack tests, (vii) sampling and laboratory characterisation of mortars, plasters and stones through chemical, physical and mechanical tests.

Complexes of historical buildings have generally been subjected to the addition of several volumes in different times, and the possible discontinuities between the different volumes could affect the overall seismic behaviour. In fig. 6, block 69 in Acuti street is shown as an example. Its volume was apparently rather simple when looked from the outside, whereas required to be subdivided into three macro-elements once surveyed in more detail (figs. 7–9). For a reliable interpretation of signs of damage, the preliminary in-situ survey, useful to obtain details of the geometry of the structures, identifying irregularities such as vertical deviations and rotations, needed complementary indication coming from the investigation of the historical evolution of the structure in its complexity (fig. 10).

On the chosen case studies, significant walls were recognized for a complete characterization of the masonry where the following tests were systematically carried out: (i) sonic tests by transparency on a grid of 75×75 cm^2 for

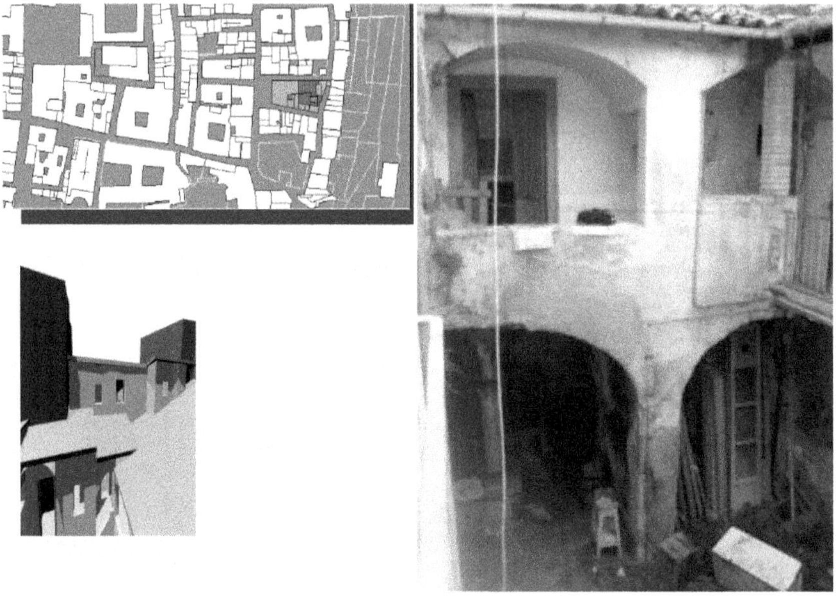

Figure 6: Location and views of block 69 in Acuti street.

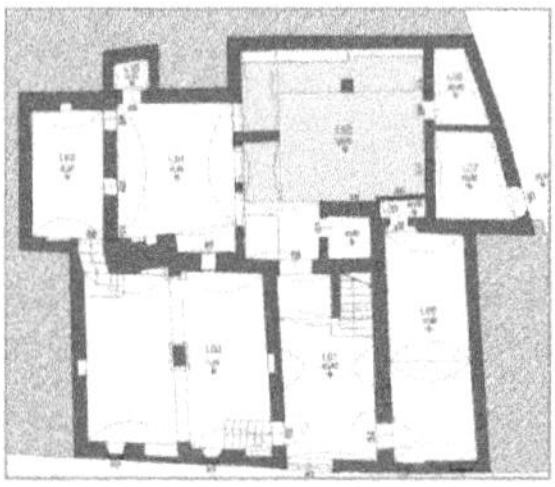

Figure 7: Ground floor of block 69.

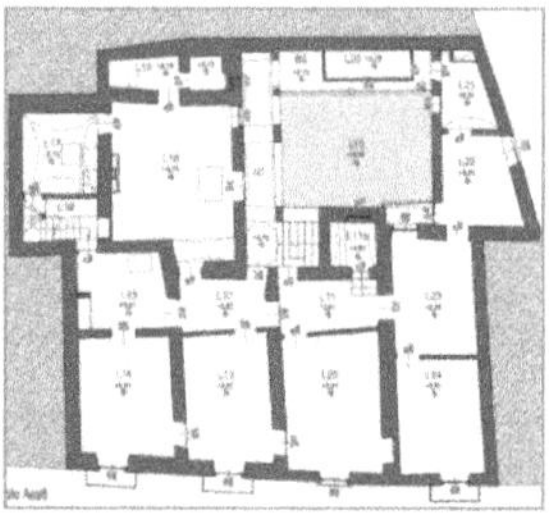

Figure 8: First floor of block 69.

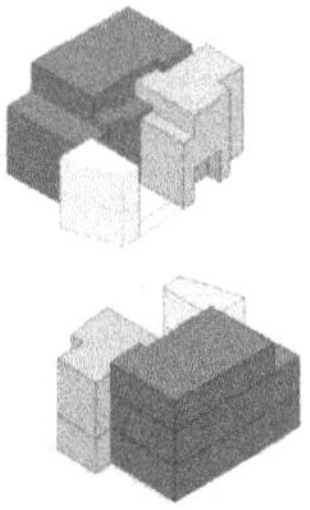

Figure 9: Macro-elements.

measurement of the sonic velocity, (ii) single and double flat-jack tests for measurement of the state of stress, the modulus of elasticity E and the coefficient of lateral expansion, (iii) survey of the masonry morphology and material sampling for identifying the chemical, physical and mechanical properties of mortar and stones, (iv) repositioning of the stones in the wall.

In fig. 11, the results of sonic tests on a pillar rising at ground floor of block 69 are reported. It is interesting to observe the agreement between the distribution of sonic velocity, with higher values corresponding to the outer portions of the pillar and lower values at mid span, and the crack pattern, showing cracks running where the lower velocity has been recorded.

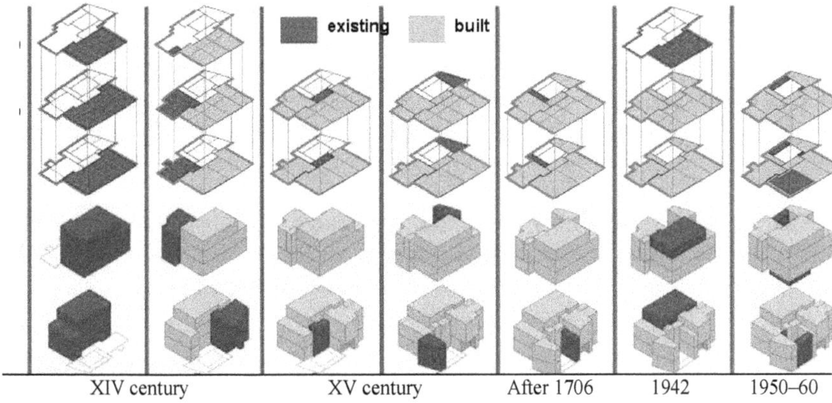

Figure 10: Historical evolution of block 69.

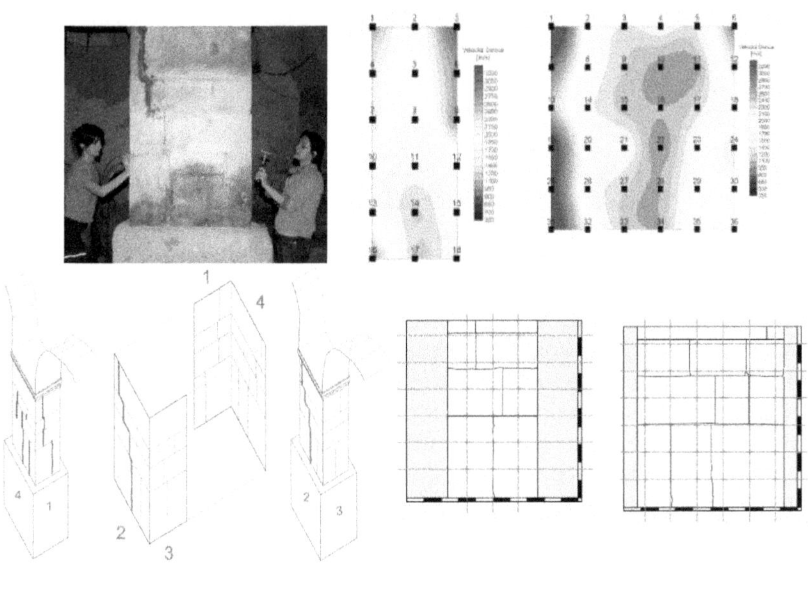

Figure 11: Sonic characterization and study of the crack pattern of a pillar at ground floor of block 69.

The survey of the morphology of the wall cross section was aimed at understanding whether the masonry was made of one or more leaves and whether the leaves were connected in some way. It was carried out by drilling, taking off some stones in order to visually investigate the wall texture, sketching the inner aspect of the wall and sampling stones and mortars for laboratory testing.

The mortars were submitted to chemical and physical analyses. The binder was separated from the aggregate by thermal attack and the grain size distribution was measured as shown in fig. 18. Since the aggregate was mainly calcareous, it was not possible to determine the binder/aggregate ratio chemically. Most of the mortars were similar, with the binder being hydrated lime. The stones belonged to several litho types but limestone was the most frequent one.

After in situ and laboratory characterization and a 3D evaluation of the crack pattern, the main weakness characteristics towards the building seismic response were singled out. In fig. 12 the presence of a standing out body and a room with the lack of the floor are indicated as an example.

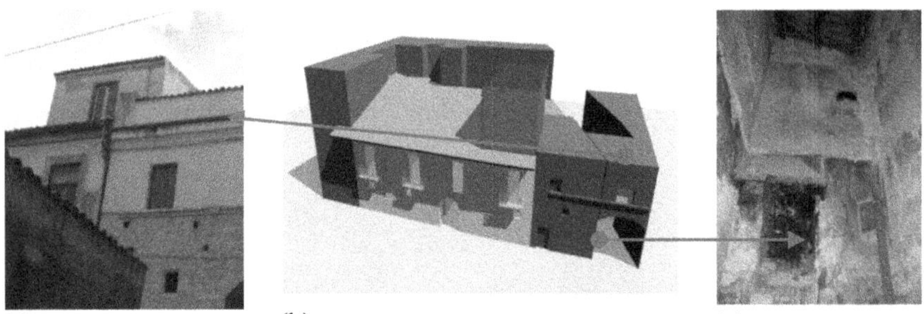

(a) (b) (c)

Figure 12: Weaknesses in block 69: standing out body (a); lack of floor (c).

Another case study is presented in fig. 13, where on the plan of the building complex different colours indicate two construction phases and capital letters refer to the testing points. With regard to point A, where parts of the building erected in different years converge, the masonry inspection after plaster removal highlighted that stone interlocking was absent (fig. 14) and therefore that the two walls presented no connections. Results of sonic and flat jack characterization of a wall located in the more recent part of the complex and of another wall located in the original part are reported in figs. 15 and 16 respectively. The results of in situ testing gave a confirmation of what was observed by visual inspection. The resulting masonry quality was generally poor, characterized by low values of sonic velocity; in some cases flat jack tests were applied with difficulty, and indicated very high values of both vertical and horizontal displacements. Fig. 17 shows the relationship between E modulus obtained through double flat jack test and sonic velocity obtained from the same masonry area, measured on different stone masonry walls from three different Italian historic centres.

4 Discussion of the results and conclusions

The seismic vulnerability assessment of historical buildings should consist of an interactive procedure based on indirect and direct sources of information, leading to the identification of the most typical failure mechanisms activated by the earthquake and allowing for appropriate analytical models. The paper describes

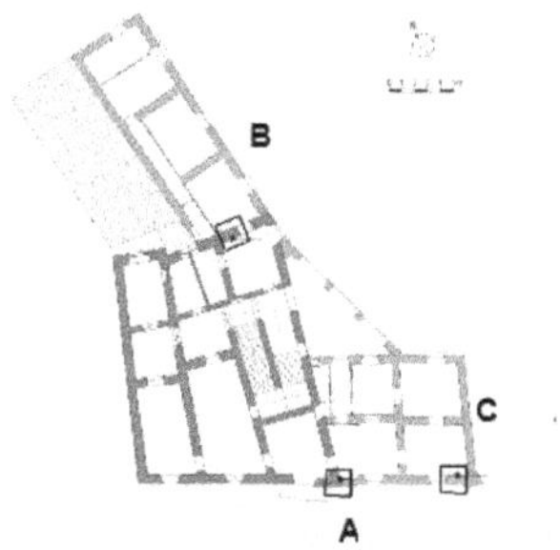

Figure 13: Sardi Palace in Palizze street: ▪ original parts; ▪ parts added in 19th century.

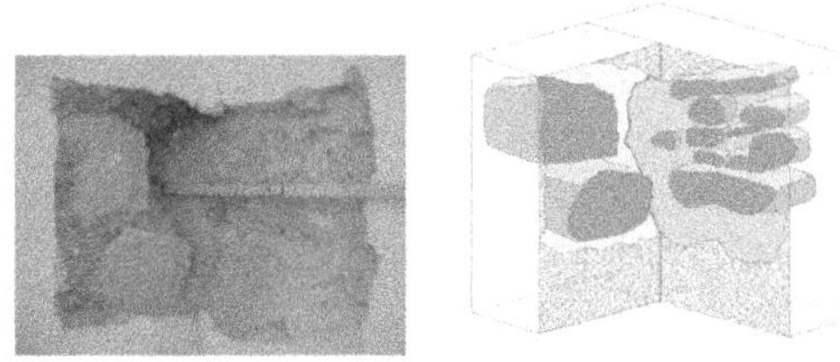

Figure 14: Study of wall interlocking at corner A of fig. 13.

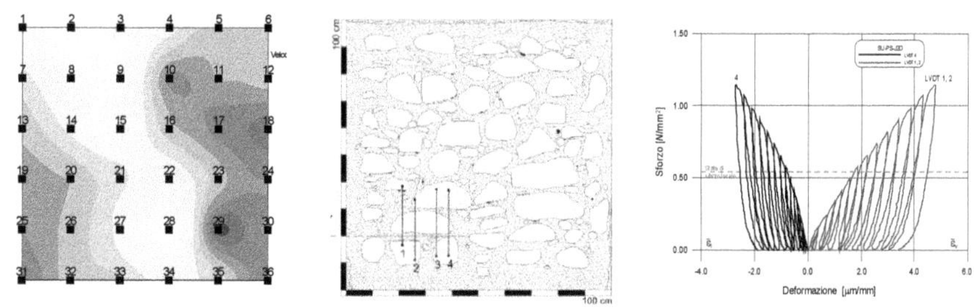

(a) *average velocity 1016.3 m/s*

(b)

(c) *average E 282 (N/mm^2)*

Figure 15: Sonic (a) and double flat jack (c) characterization of wall C (b) of fig. 13.

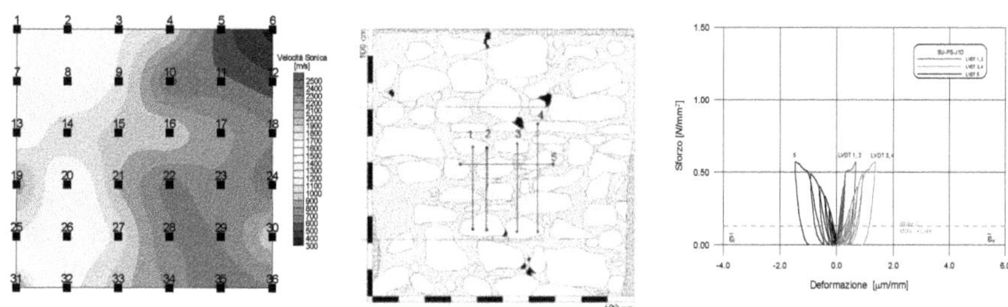

(a) *average velocity 938.6 m/s*

(b)

(c) *average E 1737- 574 (N/mm^2)*

Figure 16: Sonic (a) and double flat jack (c) characterization of wall B (b) of fig. 13.

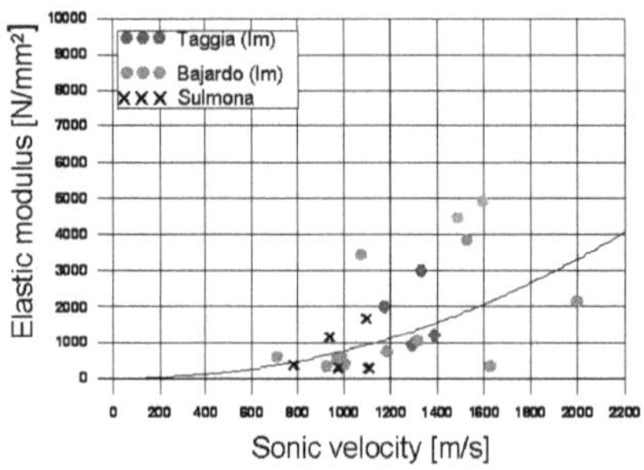

Figure 17: Relationship between *E* measured through double flat jack test and sonic velocity.

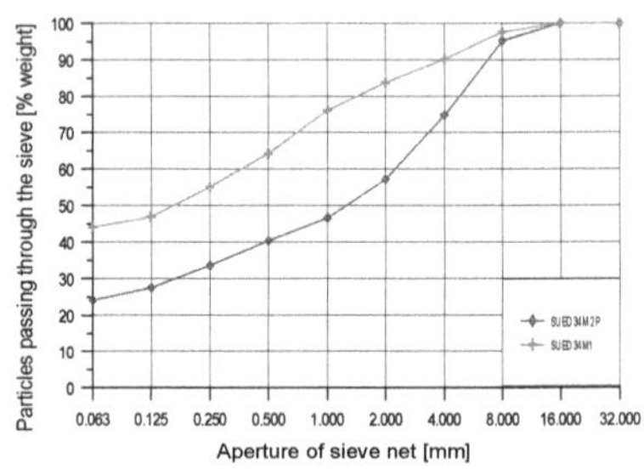

Figure 18: Grain size distribution of mortar aggregates.

the application of a multi-disciplinary procedure, addressed to an inter-linked knowledge of morphological and constructive aspects of the masonry elements of stone-masonry buildings in the historical centre of Sulmona. The results have been used for the application of the automatic procedure Vulnus version 4.0 [9] to obtain a global seismic vulnerability assessment, based on a probabilistic evaluation of the percentage of structural units exceeding a fixed damage level.

According to the experimental results, indicating a low quality of the masonry constituting the analyzed case study, it turned out that the structural units in the historical centre of Sulmona, if subjected to seismic action, show a better in plane than out of plane response, demonstrating a very high vulnerability.

In general terms, a vulnerability analysis should be carried out by examining the experimental results in the light of the historical evolution of masonry buildings, in these cases subjected to many heavy seismic events over centuries; still, more research should be carried out on mortar contribution to the general behaviour.

Acknowledgements

The authors wish to thank eng. A. Martinelli of C.N.R. I.T.C. L'Aquila and the technical support of arch. B. Matticoli. M. Antico, M. Cucchi, M. Iscandri, A. Colombo is gratefully acknowledged. Many thanks are also due to the students: C. Ferrario, S. Francotti, G. Germano, P. Lombardo, L. Restelli, D. Rossini and V. Sgobba. The research was co-financed by the National projects RELUIS supported by the Civil Protection Agency, Italy.

References

[1] OPCM 3274/2003.
[2] Binda, L., Baronio, G., Gambarotta, L., Lagomarsino, S. & Modena, C., Masonry constructions in seismic areas of central Italy: a multi-level

approach to conservation. *8th North American Masonry Conference*, Austin, USA, pp. 44-55, CD-ROM, 1999.

[3] Anzani, A., Baila, A., Penazzi, D. & Binda, L., Vulnerability study in seismic areas: the role of on-site and archives investigation. *IV International Seminar on Structural Analysis of Historical Constructions*, Vol. 2, Padova, pp. 1051-1059, 2004.

[4] Anzani, A., Binda, L., Cantini, L. Cardani, G., Saisi, A. & Tedeschi C., On site and laboratory investigation to assess material and structural damage on some churches hit by an earthquake. *XII Conv. Naz. L'Ingegneria Sismica*, ANIDIS 2007, Pisa, DVD, Edizioni Plus, Pisa, pp. 1-10, 2007.

[5] Cardani, G., Anzani, A, Binda, L. & Saisi, A., On site structural assessment of churches in stone work damaged by the earthquake. *2nd Canadian Conference on Effective Design of Structures*, CD-ROM ISBN: 0-9738035-3-2, McMaster University, Hamilton, Ontario, Canada, pp. 1-10, 2008.

[6] Penazzi, D., Valluzzi, M.R., Cardani, G., Binda, L., Baronio, G. & Modena, C., Behavior of historic masonry buildings in seismic areas: lessons learned from the Umbria-Marche earthquake. *12 IBMaC*, Madrid, 2000.

[7] W. Hendry, Aspects of stability and strength of stone masonry structures. *Third International Masonry Conference*, Proceedings of the British Masonry Society, No. 6, March, 1994.

[8] OPCM2005, Ordinanza del Presidente del Consiglio dei Ministri n. 3431 (2005). "Ulteriori modifiche ed integrazioni all'Ordinanza del Presidente del Consiglio dei Ministri n. 3274" – in Italian.

[9] Binda, L., Cardani, C., Saisi, A. Valluzzi, M.R., Munari, M. & Modena, C., Multilevel approach to the vulnerability analysis of historic buildings in seismic areas, Part 1: Detection of parameters for vulnerability analysis through on site and laboratory investigations. *International Journal of Restoration Building and Monuments*, **13(6)**, pp. 413-426, 2007.

Evaluation of the structural behaviour of historic masonry buildings by a sonic pulse velocity method

F. Casarin, M. R. Valluzzi, F. da Porto & C. Modena
Department of Structural and Transportation Engineering, University of Padova, Italy

Abstract

Sonic pulse velocity tests are mainly used in historic masonry structures for consistent diagnosis evaluations and checking of the effectiveness of strengthening interventions. In this paper, results obtained from sonic tests applied on two historical constructions, namely the cathedral of Reggio Emilia and the bell-tower of the S. Zeno basilica in Verona, Italy are presented, to allow structural models to be calibrated on an experimental basis for the assessment of the behaviour of the buildings. The sonic investigation method is preliminarily discussed in the light of the outcomes of a comprehensive EU research project, also aimed at the identification of capability and limits in applications in the historical heritage field.
Keywords: non destructive techniques, experimental investigation, historical masonry structures, FE modelling, sonic pulse velocity test.

1 Introduction

The sonic pulse velocity method is a rather established technique for the in-situ diagnosis and for the evaluation of the effectiveness of the interventions applied to historical masonry buildings [1]. It is a fully non-destructive technique, thus it does not present any risk of damage for the building.

Sonic waves are used for preliminary investigations in masonry to identify internal cohesion among materials, in order to deduce the presence of voids or damage, thus obtaining important indications for the possible most suitable consolidation intervention (for instance, injectability checking of the wall).

By extension to some data-base and studies available on non destructive test (NDT) results systematization, and by comparison with results obtained by the concomitant application of other NDT (e.g., radar, endoscopies, thermography) or minor destructive tests (MDT) (e.g., flat jack tests, coring), different typologies of masonry (stone, clay bricks), quality of materials (e.g. rubble, regular, inhomogeneous combinations) and specific morphologic or constructive aspects (e.g., multi-leaf sections, soundness of intersections), can be qualitatively identified. In contrast, due to the large variability of masonry materials, quantitative results are rather difficult to obtain in historic structures [4]; in such a field, only the direct comparison among velocities acquired in different portions of the wall investigated under the same conditions (e.g., to assess the current state of masonry for diagnosis purpose), or of the same portion under different conditions (e.g., before and after a consolidation intervention, to verify its effectiveness) can the produced results considered quantitative.

Although several investigations have been carried out on this subject, it is still possible to confirm the unreliability of absolute correlations between only the velocity and a specific identification of masonry and/or of a state of consistency of it.

Sonic test results can therefore be usefully employed to improve structural characterizations by numerical modelling (aimed at identifying the mechanical behaviour of buildings or, more locally, at component level), but whose reliability is often very low for masonry constructions, especially in damaged conditions.

In the paper, various test configurations (direct, tomography) are applied to two historical constructions in Italy, namely the Cathedral of Reggio Emilia and the bell-tower of the S. Zeno church in Verona, to calibrate structural models for the identification of their mechanical behaviour.

An in-depth investigation of the several aspects involved in the application of NDT and MDT techniques (equipment, test configurations, results processing, comparison between different techniques) was performed in the framework of the European funded project "On-Site Investigation Techniques for the Structural Evaluation of Historic Masonry Buildings" – ONSITEFORMASONRY – [7], which compiled in-situ and laboratory applications to several historic masonry structures located in different parts of Europe. The main issues of the project are summarised here, with particular attention to the capability and limits of test applications in historical sites.

2 Methodology, applicability and limitations

The sonic pulse velocity method is applied for measuring the transit time of sonic pulses (stress waves) in masonry structures. This transit time and the calculated sonic pulse velocity result in basic information about the quality and consistency of the masonry element under investigation. The velocity is influenced by the composition of the masonry as well as by the presence of inhomogeneities, voids and deteriorated areas [4].

Some published work shows that a certain relationship can be found between the sonic pulse velocity and the modulus of elasticity of the masonry [3]. However, this relationship varies according to the given masonry typology and texture, and it has to be stressed that statistical databases for rough estimations of elastic properties from velocity values are still at a very experimental phase.

The sonic pulse velocity test can be carried out in different ways, according to the different transmission methods. The direct transmission method for sonic waves involves passing of a stress wave through the thickness of the masonry wall. The initiation and reception points of the stress waves are in line with each other, on opposite sides of the masonry element. The sonic pulse velocity detected is thus affected by the quality and consistency of the masonry wall section. It is also possible to carry out sonic tomographies, where the measurement of sonic pulse velocities are combined along different ray-paths on a cross section of masonry, and are subsequently processed in order to define mean values of velocity on each portion of the wall section itself.

The chosen testing configuration depends on the matter of the investigation, such as, for example, the detection of masonry inhomogeneities (e.g., variation of masonry texture, repair interventions, presence of different materials), detection of multiple leaves and measurement of the thickness of each leaf, detection of detached external leaves, detection of voids or chimney flues, evaluation of effectiveness of repair interventions, detection of damaged portions of masonry or of crack patterns.

The frequency associated with the sonic tests (20–20,000 Hz), together with the wavelength affects the minimum size flaw that can affect the transmission of a sonic pulse [2]. The method then fails when its resolution is not sufficient for the aims of the investigation, and this can happen if detection of small voids or small inclusions is required, or in some cases when dealing with the detection of multiple leaves, detached external leaves, cracks, etc.

3 Applications to historical masonry structures

The Sonic Pulse Velocity Method was applied in combination with other techniques for the acquisition of important structural information on several masonry buildings, aiming at the subsequent assessment by means of structural models. Under the activities of the EU funded project "Improving the Seismic Resistance of Cultural Heritage Buildings [8], the cathedral of Reggio Emilia, Italy, underwent an extensive investigation campaign, including sonic tests [5]. Subsequently the structural behaviour of the building was simulated by means of analytical and numerical models.

The second case study is the bell-tower of the S. Zeno basilica in Verona, Italy. In this case, the application of sonic tests in tomographic configuration was considered for the evaluation of the inner composition of the belfry "L" shaped pillars, as a preventive evaluation related to a strengthening intervention of the roof and bells' supporting structure of the tower [6].

3.1 Santa Maria Assunta cathedral, Reggio Emilia, Italy

The origins of the Reggio Emilia cathedral are to be found in the middle of the IXth century. The structure was based on the pre-existences of an early Christian church. The original plan of the church corresponds to a "Latin cross" typology, with three naves and transept. Throughout the centuries the cathedral underwent several interventions that substantially altered the original construction. The principal modifications corresponded to: the extension of an atrium to the original façade, with a superimposed dome lantern (XIIIth century); the erection of chapels along the aisles, between the mid XVth century and the end of the XVIth, and the raising of the dome lantern (1451); the raising of the aisles' vaults, the elimination of the Romanesque women's gallery (1551–1559) and the inclusion of the original Romanesque columns inside "new" massive cruciform pillars; the substitution of the ancient dome lantern positioned at the nave/transept crossing with a dome (1624–1626), and the complete substitution of the pre-existing cross-vaulted structures, both in the central nave and in the aisles, with the construction of barrelled vaults (1777).

The cathedral presents an original Latin cross plan, with central nave, two aisles and transept. The length of the church is 77.40 m, the width is 33.80 m, the span of nave and aisles is 10.15 m and 6.50 m, respectively. The maximum height of 44.60 m is reached at the top of the dome; the height of the façade dome lantern is 33.80 m and the height of the roof above the central nave corresponds to 22.25 m. The cathedral is mainly composed by clay brickwork masonry. The diverse construction phases brought about the use of different materials, and some structural elements or parts (e.g. Romanesque pillars, lower façade veneers) are made of stone (fig. 1).

Sonic tests were extensively carried out on the Cathedral's masonry structures (2005). The main aim of the sonic investigation campaign was to qualitatively

Figure 1: Santa Maria Assunta cathedral in the city centre of Reggio Emilia.

define the masonry conditions at the tested points, in order to proceed with the subsequent phase of structural assessment.

As a starting point, a sonic direct test was performed in correspondence of the masonry pillar previously tested by means of double flat-jack. Considering the several structural interventions that modified the original layout of the religious complex, the resulting changes in materials and constructive techniques, any correlation between data emerged from sonic tests and variation of elasticity modulus was not established, since the procedure was not considered reliable. The locations of sonic tests are given in fig. 2.

The sonic tests were aimed at the qualitative assessment of the inner composition of the analyzed structural elements, since all of the tests were carried out in direct configuration. Data emerged from the application of the method to wide portions of the complex appeared globally fair velocity results, indicating possible fair mechanical characteristics. In terms of minimum values encountered, seldom the velocity was lower than 800 m/s.

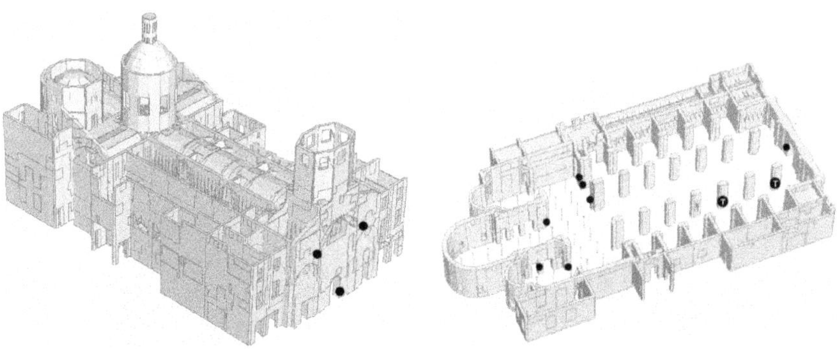

Figure 2: Reggio Emilia cathedral, locations of sonic tests.

No odd velocity values were found, indicating absence of macroscopic inconsistencies within the tested masonry portions investigated. Several masonry walls tested exhibited general velocity uniformity, indicating a possible uniformity also in terms of mechanical characteristics, resulting in a positive homogeneous stress pattern. This was particularly evident in tests S-F-H1/H2, where a façade horizontal slice 17 m long was analyzed at the same height, yielding rather uniform velocity values (fig. 3).

A significant data scatter was noticed corresponding to areas subjected to past restoration interventions, as is the case of the façade lantern back wall. Sonic results obtained in the lower part of the façade (test S-F-L) are indicative of a possible masonry variable inner configuration. Sonic velocities obtained in the upper façade tests (S-F-H1/H2) had an overall value of 1215 m/s, averaged over 61 paths, in line with data emerged from the upper lines of test S-F-L. In some cases it was possible to appreciate that the analyzed structure represents a multi leaf structure, this being indicated by the marked difference in velocity recorded in the superfacial path relative to paths crossing the inner section of the analyzed wall/pillar (fig. 4: tests S-C-FJ, S-N-D1/D2). Confirmation of this comes from the examination of the core samples extracted in the crypt area.

Figure 3: Positioning of sonic direct tests, façade (elevation) and dome lantern back wall (plan view).

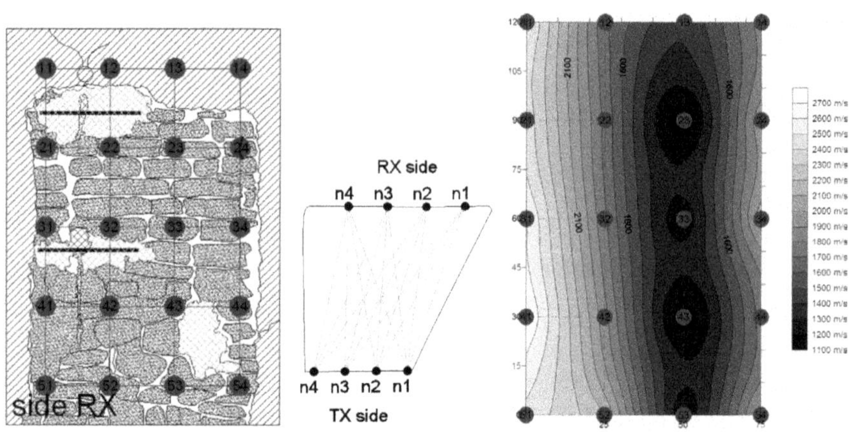

Figure 4: Reggio Emilia cathedral, crypt level: direct sonic test, acquisition grid, sonic paths and obtained velocities.

Sonic investigation results, related to the brickwork masonry typology, suggested the final impression (overall average velocity 1394 m/s from 349 sonic paths) that the subsequent implementation of structural models relying on the limit analysis approach, considering the selected masonry portions (macroelements) as rigid bodies, was possible, since the tested masonry portions did not indicate severe structural deficiencies, possible cause of sudden collapse (e.g. masonry crumbling) before the beginning of the considered mechanisms.

3.2 San Zeno bell-tower

The S. Zeno basilica is one of the most important and well known churches of Verona, being dedicated to the patron saint of the city. The origins of the early settlements of S. Zeno are related to the church and coenoby erected on the roman and early Christian graveyard area, close by the via Gallica, built on the burial place of the bishop Zeno (362–380 AD) to keep its memory and holy relics. The original buildings went through reconstructions during the VIth century, and in 805–6 AD the church was widened and a monastery was built close by.

After the destruction perpetrated in 963 AD by the Hungars, the church was built again (983 AD), with three naves, three apses and crypt, reaching its definitive width. At the end of the XI c. the church was extended and renewed with a Romanesque style. Interventions continued along the XIIth and XIIIth centuries (further extensions, raising of the façade, creation of the rose window), and in 1387 the apse and roof were demolished and rebuilt with Gothic features (fig. 5).

On the right side of the basilica and separated from it, the bell-tower rises to a height of 66 m. Erected in 1045 AD and repaired in 1120 (after the 1117 earthquake), was finished in 1178. Above the base it presents the characteristic alternation of colours due to the sequential use of calc-tufa stones and brickwork bands. The rhythm of the bell-tower is marked by cornices decorated with tuff stone arches and it is crowned by a double order of thee mullioned windows and by a spire with 4 side pinnacles.

Figure 5: S. Zeno basilica, façade. In the background, on the right, the bell-tower.

As a support for the strengthening intervention successively carried out on the S. Zeno bell-tower (2006), an in-depth study on the conditions of the belfry's structural elements was carried out, with particular focus on the masonry pillars sustaining the massive spire, whose degree of structural decay was outlined by preventive analyses (deterioration of the mortar joints caused by the rain and ice-thaw cycles, evident crack pattern, interaction with the structure supporting the bells).

In particular, sonic tomographies at two different levels of an "L" shaped pillar of the upper level of the belfry (fig. 6) were applied (at 0.55 and 1.85 m from the belfry floor) to obtain an insight on the inner composition of the masonry structural elements. Both tests indicate that the inner part of the pillar is composed by a masonry with supposed reduced masonry consistency/different composition relative to the external area, this being indicated by the different velocities emerged from the analysis (average velocity of the inner and outer areas respectively equal to approximately 1800 and 2500 m/s in the lower cross section and 1600 and 2600 m/s in the higher).

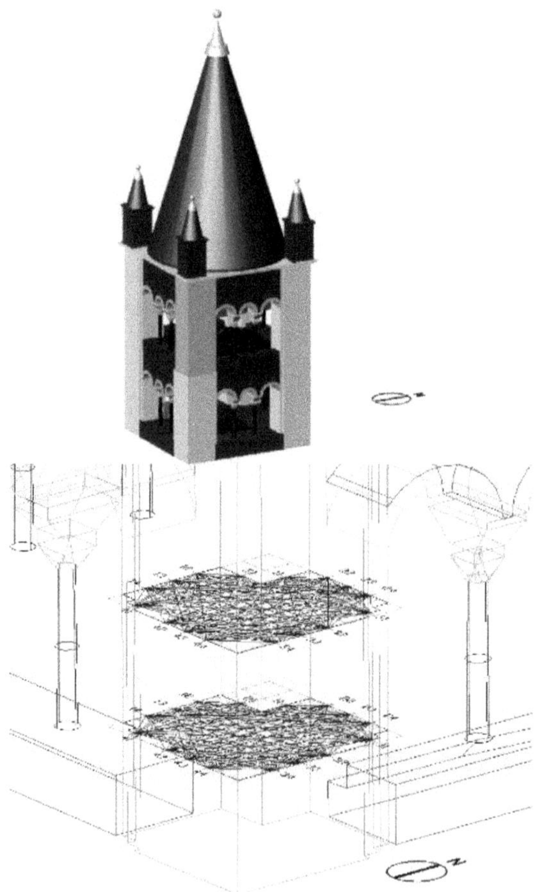

Figure 6: S. Zeno bell-tower: localization of the pillar interested by sonic testing and tomographic sections with indication of the sonic paths.

Such results, corroborated by the successive extraction of core samples in the area of execution of the tests, can be related to the presence of brickwork masonry in the inner part, resulting in a velocity that generally indicates a material with fair/good mechanical characteristics, and an outer masonry composed by well arranged sandstone masonry blocks (fig. 7). Subsequently to the outcomes of the investigation phase, structural finite element (FE) linear models of the investigated pillar, simulating the dead load and thrust of the spire, as well as the horizontal action of the swinging bells, and considering different materials throughout the cross section as emerged, were implemented. From the numerical analysis it emerges that the pillars are subjected to compressive stresses ranging from approx. 2.0 (in its external sides – respect the belfry – and inner core) to 3.0 MPa (in the sides towards the inner belfry), these evaluations giving the possibility to determine the safety state of the pillar and influencing the successive phase of strengthening intervention planning (fig. 8).

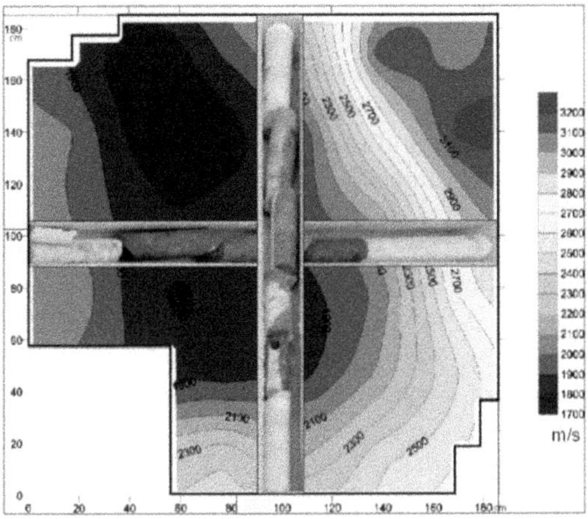

Figure 7: Lower sonic tomography, S. Zeno bell-tower, Verona: velocities are related to masonry composition by means of core samples extraction.

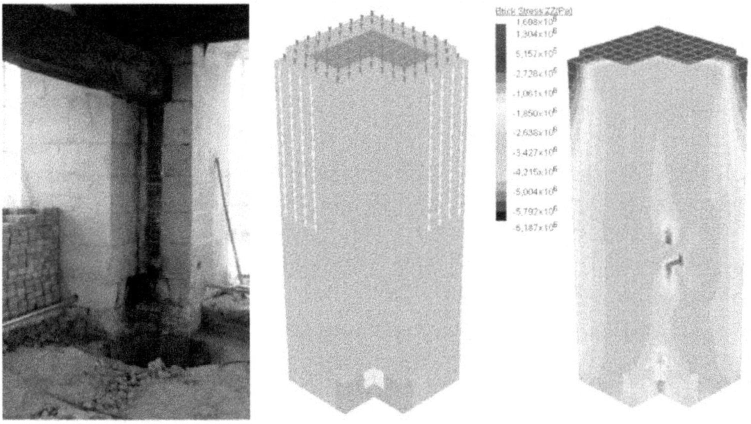

Figure 8: View of pillars of St. Zeno tower belfry and corresponding pre and post processed FE model.

4 Conclusions

From the experience acquired, it emerged that the sole application of sonic tests, considering their characteristics as ND tests, does not allow a confident implementation of structural models, in the sense that several issues related to mechanical evaluation and assessment cannot be resolved just by the identification of the sonic velocity within the structural elements of a historical masonry building.

More information may arise from the evaluation of the sonic tomography, whose velocity detection may be indicative of different typology masonry thus also with different mechanical parameters. Also, it emerged that NDT methodologies such as sonic tests have possibly to be complemented with other MDT techniques necessary for a results generalisation in order to proceed with the subsequent phase of structural assessment.

Finally, it can be stated that the most appropriate testing methodology is strongly dependent on the expected results of the acquisition campaign. In the case of the Reggio Emilia cathedral, a wide application of the sonic method (direct configuration) was necessary for the general implementation of limit analysis models used to identify the seismic response of the building. In the case of the S. Zeno bell-tower in Verona, a more detailed analysis (sonic tomography) was focused on a single structural element to solve a local FE modelling problem.

Acknowledgements

The research reported in this paper was mainly carried out under the activities of the EU funded projects “Improving the Seismic Resistance of Cultural Heritage Buildings” [7] and “ONSITEFORMASONRY” [8]. The authors would like to gratefully thank engineers Davide Semenzato and Matteo Scodeller for their collaboration for on-site tests and data processing.

References

[1] RILEM Recommendation TC 127-MS. MS.D.1 Measurement of mechanical pulse velocity for masonry, *Materials and Structures*, **29**, pp. 463-466, 1996.

[2] Epperson, G.S. & Abrams, D.P., Non destructive evaluation of masonry buildings, Advanced Construction Technology Center, Doc. N. 89-26-03, Urbana Illinois, 208 pp., 1989.

[3] Riva, G., Bettio, C. & Modena, C., Valutazioni quantitative di caratteristiche meccaniche di muratura in pietra esistenti mediante prove non distruttive. *Materiali e Strutture*, L’ERMA di Bretschneider Ed., n. 1, 1998 (in Italian).

[4] Deliverable D11.1 (2004): Technical guidelines for an appropriate use of the suggested equipment. ONSITEFORMASONRY, Project n° EVK4-2001-00091, Contract n° EVK4-CT-2001-00060, 2004.

[5] Semenzato, D. Indagini strutturali e verifica della vulnerabilità sismica della chiesa di S. Maria Assunta a Reggio Emilia. Graduation thesis, Tutor Prof. MR. Valluzzi, University of Padua, 2006 (in Italian).
[6] Scodeller, M., Indagini soniche per la diagnosi di strutture murarie a torre. Graduation thesis, Tutor Prof. MR. Valluzzi, University of Padua, 2006 (in Italian).
[7] EU funded research project (5th Framework Programme) "On-site investigation techniques for the structural evaluation of historic masonry buildings, Project Acronym: ONSITEFORMASONRY, Project Reference: EVK4-CT-2001-00060, http://www.onsiteformasonry.bam.de.
[8] EU funded research project (EU-India ECC Programme) "Improving the Seismic Resistance of Cultural Heritage Buildings", Contract ALA-95-23-2003-077-122, http://www.civil.uminho.pt/eu-india.

Earthquake resistance of a historical brick building in Akita Prefecture, Japan

C. Cuadra, K. Tokeshi, M. B. Karkee & Y. Sakaida
Department of Architecture and Environment Systems, Akita Prefectural University, Tsuchiya, Japan

Abstract

During the Meiji period (by the end of the 19th century) many brick masonry buildings were constructed in Japan. However, due to the lack of reinforcement, these structures collapsed during the great Kanto earthquake, which occurred in the year 1923. Since then, this type of structure is not used for buildings and only a few historical constructions remain from that era. In Akita Prefecture, in the northeastern part of Japan, some of these buildings have been declared as local culture heritages. One of them is located in Ani village and was constructed in 1879 to serve as the residence for a German engineer who was working for a local mining company. As an initial step to evaluate the seismic vulnerability of this historical building, dynamic characterization has been undertaken. For that purpose, measurements of the micro vibration of the building was planned and undertaken by the authors. For the analytical modeling, mechanical parameters were estimated from a series of laboratory tests on masonry brick units obtained from the stock that are conserved near the building under study. The results of the microtremor measurements are discussed in relation to the analytical procedure adopted to estimate the dynamic characteristics of this historical brick structure. The structure shows intricate modes of vibration that are reflected in the multiple peaks observed in the transfer functions of microtremor records. With the analytical simulation considering only the effect of the brick walls, the multiple predominant frequencies are obtained for a certain range. With this good agreement of the analytical procedure and the measurement results, the reliability of the employed methodology was verified.
Keywords: masonry building, dynamic behaviour, microtremor measurements, finite element method, mode analysis.

1 Introduction

Masonry buildings are used widely around the world. In Japan during the Meiji period, the number of builders that used masonry increased and the masonry structure became a common technology for construction. However, due to its poor seismic performance, especially during the great Kanto earthquake at the beginning of the 20th century, the technology was abandoned and was replaced by reinforced concrete structures. Therefore, only a few masonry constructions remain at present and have become historical constructions.

Some of these buildings are located in the northern part of Japan, specifically in Akita Prefecture, and most of them have been declared local cultural heritage. One of these buildings, the object of this study, is the house of a former mining engineer who was the superintendent in Ani village.

In recent years, each region, city or town of Japan has been active to show their particularities, especially, in history and local cultural aspects. If some masonry buildings are located in these zones, then they are included in these activities and therefore, it is necessary to consider the restoration and conservation of these buildings. For this purpose, the evaluation of their seismic behavior and vulnerability is needed. Then, if it is required, the corresponding proposal for their conservation, reparation or restoration can be made. In Japan, the study of these types of buildings is essential since there are not specific regulations for masonry, and only little research has been done in the area of masonry constructions.

In this investigation, an attempt is made to evaluate the seismic performance of a historical masonry building located in Akita Prefecture, in the northeastern part of Japan. The methodology for the seismic vulnerability evaluation is based on microtremor measurements to estimate the vibration characteristics of the building. Then, structural analysis is performed to compare the results with those obtained experimentally.

2 Description of the building object of the present study

The building, object of this study, is located in Ani village in Akita Prefecture and it was constructed during the 12th year of the Meiji period (year 1879). It was constructed to be the house of Adolf Mecker, a German engineer that came to Japan to work in the mines of the Akita region. As can be observed in fig. 1, the main structure is formed by the masonry walls and complemented by a wooden structure that forms a kind of fence around the building. The architectural style corresponds to the Gothic Renaissance with some arches to form the windows and door openings. The masonry units were fabricated with clay of the area and burned in a factory constructed for that specific purpose. The design and the direction of the construction of the house were done by Mr Mecker himself.

After finishing his employment contract, Mr Mecker returned to his country and the house was used first as a government office, and then as a social club for the mining workers.

Figure 1: Building object of the study.

The building under study, has suffered the action of some earthquakes, and the most recent was the Moriyoshiyama earthquake in the year 1982. This earthquake was a near source earthquake and partially destroyed the building. The Prefecture government reconstructed the building replacing the triangular gable wall of the second story by light material instead of clay brick masonry. Then only the first floor walls remain with the original dimensions and with clay brick lied with cement mortar.

The building was designated as prefecture cultural heritage in the year 1956, and recently in the year 1990, was recognized as Japanese national cultural heritage.

3 Measurements of the micro vibrations

Fig. 2 shows the plan distribution of the building, where the sensors were located in the following arrangement for the first set of measurements: one sensor at the first floor (1), five sensors at the second floor level (2, 3, 4, 5, 6) and one on the ground (7). For a second set of measurements, the sensors 2 and 7 were located at the top of the central beam of the roof. These points of measurement were called 2′ and 7′. The measurement of microtremors were measured in the three principal directions (east-west (EW), north-south (NS) and up-down (UD)) at each point. The micro vibrations were measured simultaneously during 500 s with a sampling frequency of 100 Hz. Then, stationary portions of the records were selected to calculate the transfer function that permits the estimation of the predominant periods of vibration of the structure. The results of the transfer function for the horizontal vibrations are shown in figs. 3 and 4. The

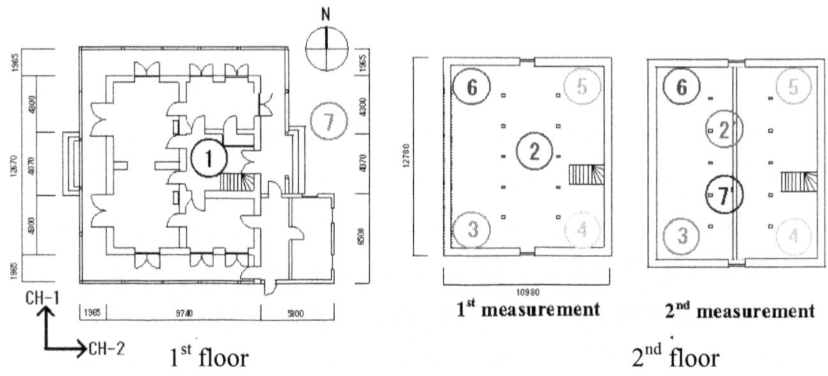

Figure 2: Points of microtremor measurements.

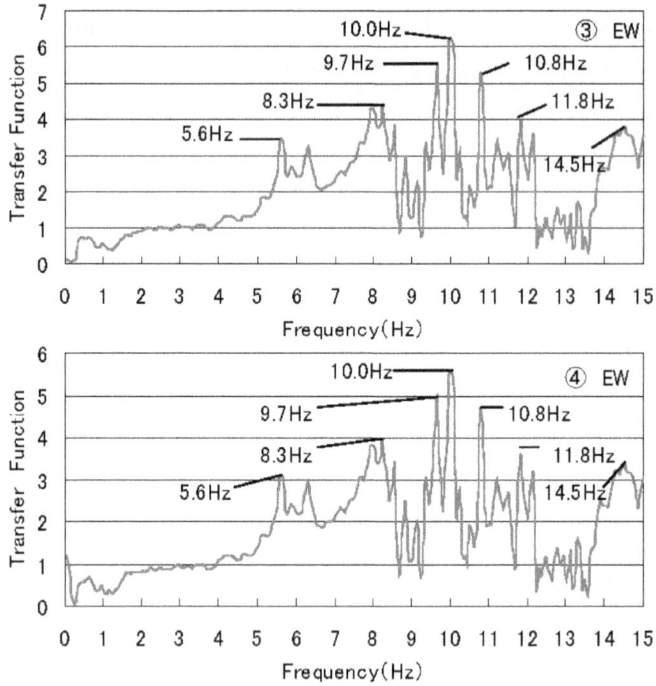

Figure 3: Transfer function results for EW direction.

transfer function at each measured point of the structure was divided by the one at the first floor level, to obtain only the vibration characteristics of the upper structure. According to these transfer function results, multiple peaks can be observed in such a way that it is not possible to obtain only one characteristic frequency. Therefore, the mode of vibration is not a simple horizontal normal mode. Instead, the modes of vibration are complex. These complex modes of vibration can be explained by the concentration of the mass of the structure in the walls instead of in the slab.

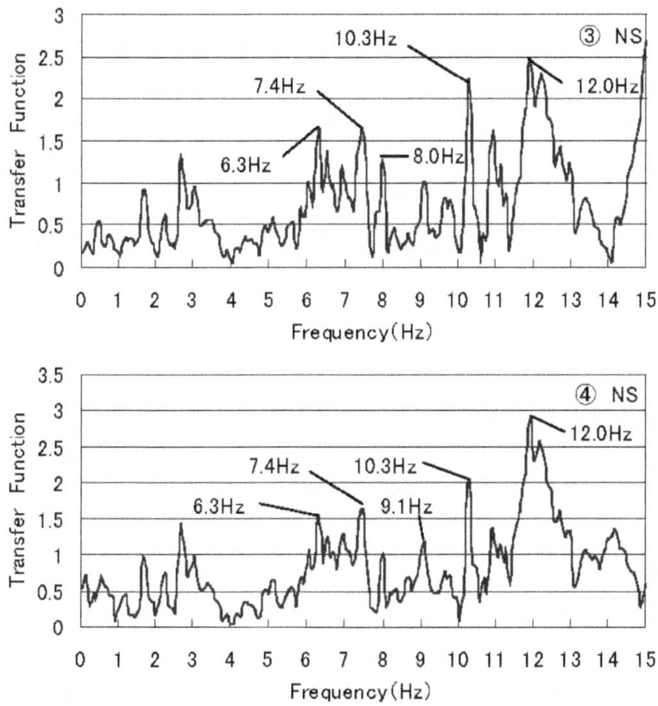

Figure 4: Transfer function results for NS direction.

4 Characteristics of brick units

Brick units selected from those that were stocked near the building were used to determine the physical and mechanical properties of the bricks. From the stock, it was difficult to distinguish between those units used in the old construction and those used for the repair works. Independently of this distinction, units showing good condition were selected for the laboratory tests. These units can be seen in fig. 5.

Figure 5: Brick units employed to obtain the physical and mechanical properties.

Water suction ratio, density, compression strength and modulus of elasticity were the properties obtained and the average values are presented in table 1.

Table 1: Physical and mechanical properties of brick units.

Suction ratio (%)	Density (gr/cm^3)	Compression strength (N/mm^2)	Young's modulus (N/mm^2)
24	1.64	12	1.06×10^4

5 Finite element modelling

In order to compare the measurement results with analytical predictions, a finite element model (FEM) was constructed considering only the brick masonry walls. This assumption is made with the intention of obtaining the influence of the wall masses on the modes of vibration. Here it is assumed that the wall, including the mortar, has homogenous properties and they are equal or similar to those obtained from the brick units. The model was built by referring to the available plans and direct measurements of the building. In total, 280 parallelepiped solid elements were used. The complete model is displayed in fig. 6.

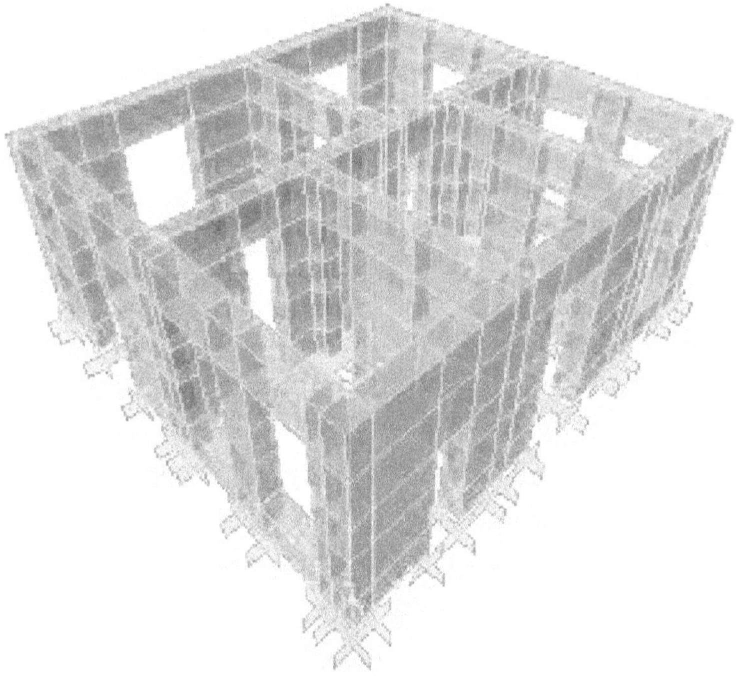

Figure 6: Finite element model.

The 10 first modes of vibration were obtained from the three-dimensional analysis in which frequencies varied from 8.3 Hz to 12.6 Hz. By comparing these with the results of the microtremor measurements, a good agreement is observed in the modes of vibration detected as peaks in the transfer functions. In tables 2 and 3, the comparison of the predominant frequencies obtained from measurements with analytical results is presented for EW and NS directions, respectively. The right column indicates the measurement point at which the predominant frequency was observed. The finite element method result shows in brackets the order of the mode of vibration obtained from the 3D analysis. As an illustrative example, the first 2 modes of vibrations are shown in fig. 7.

Table 2: Comparison of measured frequencies and FEM results (EW direction).

Microtremor measurements (Hz)	FEM analysis (Hz)	Point of measurement
8.3	8.3 (mode 1)	2’, 3, 4, 5, 6, 7’
8.8	8.8 (mode 2)	2’, 5, 6, 7’
9.2	9.3 (mode 4)	2’, 5, 6, 7’
10.8	10.7 (mode 6)	3, 4, 5, 6, 7’
11.8	11.9 (mode 8)	3, 4, 7’

Table 3: Comparison of measured frequencies and FEM results (NS direction).

Microtremor measurements (Hz)	FEM analysis (Hz)	Point of measurement
9.1	9.1 (mode 3)	3, 4, 5
10.3	10.3 (mode 5)	3, 4, 5,
10.9	11.1 (mode 7)	3, 4, 5, 7’
12.0	12.0 (mode 9)	3, 4, 5, 6
12.5	12.6 (mode 10)	2’, 3, 4, 5, 6, 7’

Figure 7: Analysis results for the first 2 modes of vibration.

6 Conclusions

From the micro vibration measurements, several predominant frequencies were obtained from the transfer functions. According to these results, it can be assumed that the building has complex modes of vibration.

To perform the analysis, first the mechanical and physical properties of the brick units were estimated from laboratory tests. These properties have a great spread in their values, but their ranges lie between the usual values for clay bricks.

The performed structural analysis was based on the properties of bricks and considering only the walls of the building, since they constitute the main part of structure mass, and therefore they have a great influence on the shape of vibration. Results show that there is a good agreement between the observed frequencies and those obtained analytically within the range of 8.3 Hz to 12.6 Hz.

Microtremor measurements combined with FEM analysis provided a valuable basis for the evaluation of the dynamic characteristics of heritage masonry structures.

References

[1] Kanai, J., Tokeshi, K., Cuadra, C. & Karkee, M.B., Vibration characteristics of buildings using microtremor measurements. *First European Conference on Earthquake Engineering and Seismology* (a joint event of the 13th ECEE and 30th General Assembly of the ESC), Geneva, Switzerland, Paper Number: 708, 2006.

[2] Cuadra, C., Sato, Y., Tokeshi, J., Kanno, H., Ogawa, J., Karkee, M. B. & Rojas, J., Evaluation of the dynamic characteristics of typical Inca heritage structures in Machupicchu. *STREMAH IX*, eds. C.A. Brebbia & A. Torpiano, WIT Press: Southampton, pp. 237-244, 2005.

[3] Ogawa, J., Cuadra, C., Karkee, M.B. & Rojas, J., A study on seismic vulnerability of Inca's constructions. *Proceedings of the 4th International Conference on Computer Simulation in Risk Analysis and Hazard Mitigation. Risk Analysis IV*, Rhodes, Greece, pp. 3-12, 2004.

[4] Cuadra, C., Karkee, M.B., Ogawa, J. & Rojas, J., Preliminary investigation of earthquake risk to Inca's architectural heritage. *Earthquake Resistant Engineering Structures IV*, eds. C.A. Brebbia & G. Latini, WIT Press: Southampton, pp. 167-176, 2003.

Assessment of masonry strength in a heritage building

M. Holicky, M. Hrabanek, J. Kolisko & M. Sykora
Klokner Institute, Czech Technical University in Prague, Czech Republic

Abstract

Heritage buildings in the Czech Republic are made of different types of masonry. Decisions concerning upgrades of these buildings should be preferably based on a reliability assessment, taking into account actual material properties. Due to inherent variability of historical masonry, information on its actual mechanical properties has to be obtained from tests. Estimation of masonry strength from measurements may then be one of the key issues in the assessment of historical structures. In this study, the standard technique provided in the Eurocode EN 1996-1-1 is applied for the assessment of a masonry structure built in the 19th century. Characteristic and design values of masonry strength, derived using principles of the Eurocode, are compared with corresponding fractiles of a developed probabilistic model. It appears that the characteristic value based on the probabilistic model is lower than that obtained by the standard technique. On the contrary, the partial factor for masonry recommended in EN 1996-1-1 seems to be rather conservative.
Keywords: masonry, characteristic strength, statistical methods.

1 Introduction

At present, heritage structures, particularly those located in urban areas, are often affected by numerous environmental influences that may cause deterioration and gradual loss of their durability and reliability. Hence protection and conservation of heritage structures, including design of adequate construction interventions, is an important issue for various experts, such as art historians, architects and civil engineers, in most European countries. Construction interventions may also become necessary due to the change of use of heritage structures. The rehabilitation of heritage structures is also a matter of great economic significance, as more than 50% of all construction activities apply to existing structures, including heritage structures. Decisions about various interventions should always be a part of the complex assessment of a heritage structure that should be based on relevant input data, including information on actual material properties.

In the Czech Republic numerous heritage structures are made of different types of masonry. Due to the inherent variability of historical masonry, information on its actual mechanical properties has to be obtained from tests. Estimation of masonry strength from measurements may then be one of key issues of the assessment of heritage structures.

This study focuses on the assessment of masonry strength of a historical structure built in the 19th century. Masonry strength is estimated from a limited number of destructive tests and a series of non-destructive tests. The standard method provided in the Eurocode EN 1996-1-1 [1] is supplemented by the use of classical statistical techniques, including the method of moments and test of outliers. The characteristics and design values of masonry strength derived using the principles of the Eurocodes are compared with the corresponding fractiles of a proposed probabilistic model.

2 Evaluation of tests

The historical residential house, located in the downtown area of Prague, was built in about 1890. The six-storey masonry building is shown in fig. 1. The structural analysis consists of models for several structural parts. In this paper, the key issue of estimation of unreinforced masonry strength is described in detail.

The mechanical properties of historical masonry are strongly dependent on the properties of its constituents. Commonly, there is a large variability of mechanical properties within a structure due to workmanship and inherent variability of materials as indicated by Lourenco [2] and Stewart and Lawrence [3]. In the present case, information about material properties needs to be obtained from tests. A series of non-destructive tests was supplemented by a few destructive tests. In addition, previous experience on the accuracy of applied testing procedures is taken into account in the evaluation of test results. The masonry of a wall and foundations is indicated in fig. 2.

2.1 Strength of masonry units

Non-destructive tests of the strength of masonry units by Schmidt hammer were made in 33 selected locations all over the structure. A histogram of the obtained measurements is indicated in fig. 3. It appears that the sample includes an extreme measurement (maximum) that may result from an error within the measurement procedure. Therefore, the test proposed by Grubbs [4] is used to indicate whether the hypothesis that there is no outlier in the sample can be rejected or not. For the significance level 0.05 the test indicates that the hypothesis can be rejected and the measurement is deleted from the sample.

Figure 1: View of the assessed building.

Figure 2: Masonry of a wall and foundations.

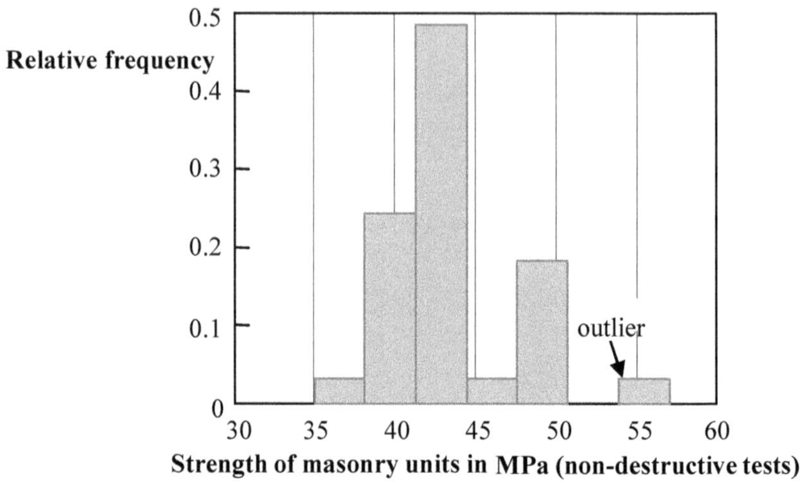

Figure 3: Histogram of the masonry unit strength obtained by non-destructive tests.

Table 1: Statistical characteristics of variables influencing the masonry strength.

Variable	Symbol	Mean	Coefficient of variation	Skewness
Strength of masonry units (non-destructive tests)	f_b	43.1 MPa	0.08	0.15
Conversion factor – masonry units	η_b	0.45	0.2	unknown
Strength of mortar (non-destructive tests)	f_m	1.26 MPa	0.41	-0.06
Conversion factor – mortar	η_m	1	0.2	unknown
Model variable	K	0.66	0.2	unknown

Point estimates of the sample characteristics – mean, coefficient of variation and skewness – are then estimated by the classical method of moments described by Ang and Tang [5], for which prior information on the type of an underlying distribution is not needed. The sample characteristics are indicated in table 1.

It appears that the sample coefficient of variation and skewness of the masonry unit strength estimated by the non-destructive tests are low. These characteristics may provide valuable information for the choice of an appropriate statistical distribution to fit the sample data. However, it is emphasized that the sample size may be too small to estimate the sample skewness with confidence.

The sample characteristics in table 1 indicate that the strength of masonry units estimated by the non-destructive tests might be described by a two-parameter lognormal distribution having the lower bound at the origin (LN0) or by a more general three-parameter shifted lognormal distribution having the lower bound different from zero (LN). Another possible theoretical model is the popular normal distribution.

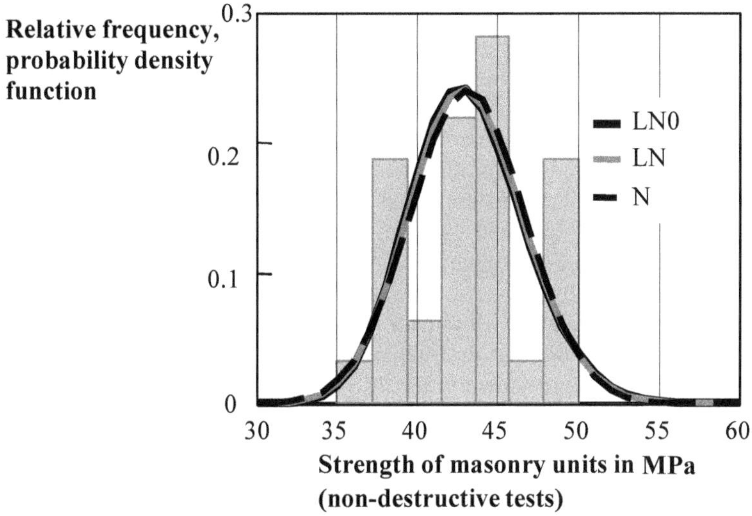

Figure 4: Histogram of the masonry unit strength obtained by non-destructive tests without the outlier and the considered theoretical models.

Probability density functions of these three theoretical models (considering sample characteristics) and a sample histogram without the outlier are shown in fig. 4. It follows that, due to the low sample coefficient of variation and skewness, all the considered models describe the sample data similarly. To compare goodness of fit of the considered distributions, Kolmogorov-Smirnov and chi-square tests described by Ang and Tang [5] are further applied. It appears that no distribution should be rejected at the 5% significance level; however, the lognormal distribution LN0 seems to be the most suitable model. Therefore, this distribution is considered hereafter.

The conversion factor η_b is further taken into account to determine the normalised compressive strength of masonry units f_b:

$$\eta_b = f_b \,/\, f_b' \qquad (1)$$

where f_b' denotes the strength of masonry units estimated from the non-destructive tests. Previous experience indicates that the coefficient of variation of the conversion factor may be assessed at around the value 0.2. Using a limited number of measurements, the mean value of the conversion factor was estimated at around the value 0.45.

2.2 Mortar strength

The estimation of mortar strength may be a complicated issue since sufficiently large specimens for destructive tests can rarely be taken. Therefore, non-destructive testing based on a relationship between hardness and strength of mortar was developed in the Klokner Institute of the Czech Technical University in Prague.

This method is used in the assessment. The histogram from 29 measurements is indicated in fig. 5. Point estimates of the sample characteristics given in table 1 are estimated using the method of moments. The sample coefficient of variation of mortar strength is considerably greater than that of the strength of masonry units. The sample distribution seems to be nearly symmetric as the skewness is around zero. This indicates that a normal distribution might be a suitable model. However, the normal distribution is not recommended for the description of the variables with the coefficient of variation exceeding, say, 0.20 as negative values can be predicted. Due to the zero skewness, a three-parameter lognormal distribution yields a similar model as the normal distribution. Therefore, the lognormal distribution LN0 is assumed hereafter for the mortar strength estimated by the non-destructive tests. Probability density functions of the theoretical models are shown in fig. 5.

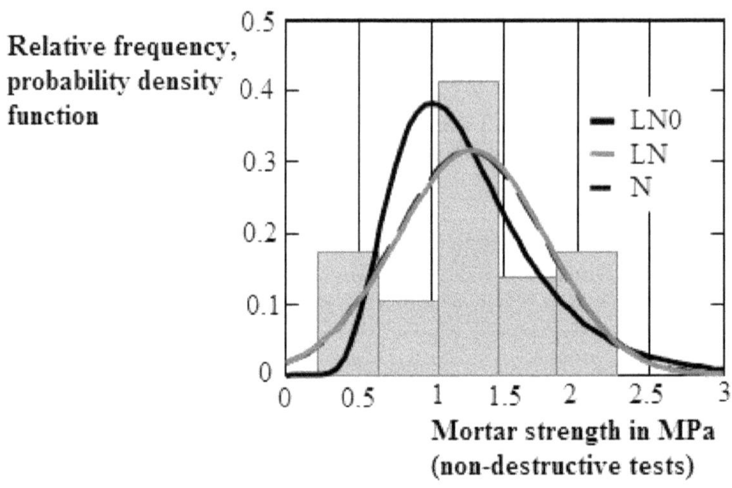

Figure 5: Histogram of the mortar strength obtained by the non-destructive tests and the considered theoretical models.

The conversion factor η_m is applied to derive the compressive strength of masonry mortar f_m from the results of the non-destructive tests:

$$\eta_m = f_m / f'_m \tag{2}$$

where f'_m is the mortar strength estimated from the non-destructive tests. Previous experience indicates that the conversion factor has the unit mean and coefficient of variation 0.2 as indicated in table 1.

3 Masonry strength in accordance with EN 1996-1-1

According to EN 1996-1-1 [1], the characteristic compressive strength of unreinforced masonry made with general purpose mortar can be estimated as:

$$f_k = Kf_b^{0.7} f_m^{0.3} = K(\mu_{\eta_b}\mu_{f'_b})^{0.7}(\mu_{\eta_m}\mu_{f'_m})^{0.3}$$
$$= 0.55\times(0.45\times43.1)^{0.7}\times(1\times1.26)^{0.3} = 4.7 \text{ MPa} \quad (3)$$

where K is the model variable and μ denotes the mean value. In the present study, Group 1 of masonry units is assumed and the model variable is 0.55. Note that a rather simplified empirical model for the masonry strength considered in EN 1996-1-1 [1] may not fit available experimental data properly. Other theoretical models may then be used to describe the compressive strength of a particular type of masonry. For instance, the application of an exponential function similar to that in eqn (3), but with general exponents, may improve the estimation of the resulting strength. More advanced models can be found in [6].

The design value of the masonry strength is derived from the characteristic value using the partial factor γ_M:

$$f_d = f_k / \gamma_M = 4.7/2.5 = 1.9 \text{ MPa} \quad (4)$$

The partial factor is dependent on a category of masonry units and classes that may be related to execution control. However, EN 1996-1-1 [1] provides insufficient guidance on classification of masonry into the proposed categories of a quality level. Following recommendations of the Czech National Annex to EN 1996-1-1 [1], the partial factor 2.5 seems to be appropriate in this case. Note that dependence of partial factors on masonry and execution control is thoroughly analysed in the study by Holicky *et al.* [7].

4 Probabilistic model

The international council ICOMOS [8] indicates that present standards and professional codes of practice adopt a conservative approach including the partial factor method to take into account various uncertainties. This may be appropriate for new structures where safety can often be easily increased. However, such an approach may fail for historical structures where requirements to improve the strength may lead to demanding repairs and loss of a cultural and heritage value.

Therefore, a probabilistic model for the masonry strength is proposed to estimate the characteristic and design values from the statistical data obtained by the tests and from previous experience and reduce the uncertainties implicitly covered by the model in EN 1996-1-1 [1]. Considering eqn (3), the compressive strength of masonry – random variable f, is given by:

$$f = K(\eta_b f'_b)^{0.7} (\eta_m f'_m)^{0.3} \qquad (5)$$

All the variables in eqn (5) are considered as random variables. Statistical characteristics are provided in table 1.

Considering background information and experimental results provided in [9], the mean of the model variable K is considered as 1.2-times the value given in EN 1996-1-1 [1] and the coefficient of variation is 0.2. Note that the variability of the model variable K describes model uncertainties and covers deviations and simplifications related to the model in EN 1996-1-1 [1].

In the previous section, the lognormal distribution LN0 is proposed to describe the variability of the strength of masonry units and mortar estimated by the non-destructive tests. In the absence of statistical data and considering general experience, the lognormal distribution LN0 is adopted also for the other variables influencing the strength of masonry. However, it is emphasized that if there is any evidence to support another distribution, then such a distribution should be preferably applied.

When all the basic variables included in eqn (5) are described by the lognormal distribution LN0, it can be easily shown that the strength of masonry has also the lognormal distribution LN0. The natural logarithm of the masonry strength is normally distributed with the mean and standard deviation:

$$\mu_{\ln(f)} = \mu_{\ln(K)} + 0.7[\mu_{\ln(\eta_b)} + \mu_{\ln(f'_b)}] + 0.3[\mu_{\ln(\eta_m)} + \mu_{\ln(f'_m)}]$$

$$\sigma_{\ln(f)} = \sqrt{\sigma^2_{\ln(K)} + 0.7^2[\sigma^2_{\ln(\eta_b)} + \sigma^2_{\ln(f'_b)}] + 0.3^2[\sigma^2_{\ln(\eta_m)} + \sigma^2_{\ln(f'_m)}]} \qquad (6)$$

where $\mu_{\ln(X)}$ and $\sigma_{\ln(X)}$ denote the mean and standard deviation of ln(X):

$$\mu_{\ln(X)} = \mu_X - 0.5\ln[1 + V_X^2]; \quad \sigma_{\ln(X)} = \sqrt{\{\ln[1 + V_X^2]\}} \qquad (7)$$

where μ_X and $V_X = \sigma_X / \mu_X$ are the mean and coefficient of a variable X given in table 1. From eqns (6) and (7), the mean 5.5 MPa and coefficient of variation 0.29 of the masonry strength are derived.

In accordance with EN 1996-1-1 [1], the characteristic strength of masonry corresponds to the 5% fractile of the assumed statistical distribution. In the present case the fractile of the lognormal distribution is 3.3 MPa. The probability density function of the masonry strength and the characteristic and design values are indicated in fig. 6. It appears that the 5% fractile of the probability distribution is by about 30% lower than the characteristic value estimated by eqn (3) that seems to be considerably non-conservative. Similar findings have been achieved earlier by Holicky *et al.* [9].

In accordance with EN 1990 [10], the design value of the masonry strength f_d is the fractile corresponding to the probability:

$$p_d = \Phi(\alpha_R \times \beta) = \Phi(-0.8 \times 3.8) = 0.0012 \qquad (8)$$

where $\Phi(\cdot)$ is the cumulative distribution function of the standardised normal distribution, the FORM sensitivity factor α_R is approximated by the value –0.8

recommended for the leading resistance variable and the target reliability index β is 3.8 for a fifty-year reference period.

The 1.2‰ fractile of the probability distribution is 2.3 MPa. Remarkably, the theoretical design value is by about 20% greater than the design value estimated by eqn (4). It follows that the assumed partial factor 2.5 may be rather conservative since, from the probability distribution, the partial factor 1.4 is estimated.

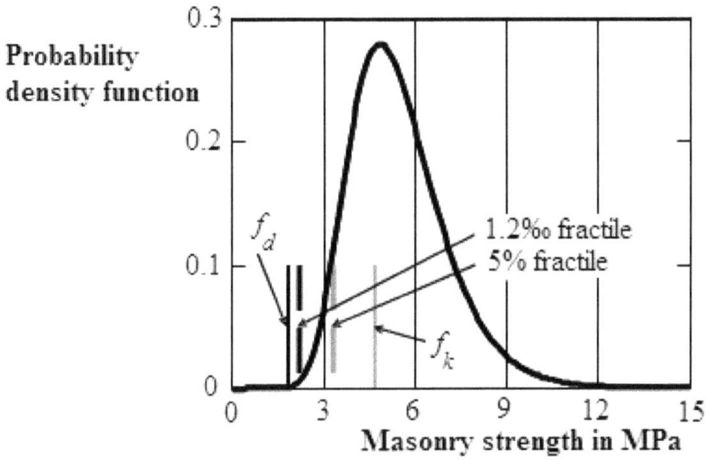

Figure 6: Probability density function of the masonry strength and the characteristic and design values.

Table 2: FORM sensitivity factors of the variables influencing the masonry strength.

Variable	Symbol	FORM sensitivity factor
Strength of masonry units (non-destructive tests)	f_b'	0.20
Conversion factor – masonry units	η_b	0.49
Strength of mortar (non-destructive tests)	f_m'	0.42
Conversion factor – mortar	η_m	0.21
Model variable	K	0.70

5 Sensitivity analysis

A sensitivity analysis is further conducted to investigate the importance of basic variables on the resulting probabilistic model. The FORM sensitivity factors given in table 2 are evaluated by the software package Comrel®.

Table 2 shows that the model variable K is the most influential variable. It follows that the proposed model may be improved particularly by reducing variability of this variable.

6 Conclusions

The following conclusions are drawn from the presented assessment of masonry strength of a historical masonry:

(1) Due to inherent variability of historical masonry, information on its actual mechanical properties has to be obtained from tests and estimation of masonry strength from measurements may be one of key issues in assessment of heritage structures.

(2) Available samples should be verified by an appropriate test of outliers as extreme measurements, possibly due to an error, may significantly affect sample characteristics.

(3) Appropriate models for basic variables influencing masonry strength should be selected on the basis of the statistical tests, taking into account general experience with distribution of masonry unit strength.

(4) The Lognormal distribution having the lower bound at the origin may be a suitable model for masonry strength.

(5) 5% fractile of a proposed probabilistic model for masonry strength is by about 30% lower than the characteristic value according to EN 1996-1-1.

(6) The partial factor 2.5 assumed in the model of EN 1996-1-1 seems to be rather conservative as compared with the partial factor 1.4 estimated from the probability distribution.

(7) The theoretical design value (1.2‰ fractile) is greater by about 20% than the design value estimated in accordance with EN 1996-1-1.

(8) The model for masonry strength may be improved particularly by reducing the variability of the model variable K.

Acknowledgements

This study has been conducted at the Klokner Institute, Czech Technical University in Prague, within the framework of the research project Assessment of Historical Immovables A/CZ0046/2/0013, supported by a grant from Iceland, Liechtenstein and Norway through the EEA Financial Mechanism and the Norwegian Financial Mechanism.

References

[1] EN 1996-1-1, Eurocode 6 – Design of masonry structures – Part 1-1: General rules for reinforced and unreinforced masonry structures, CEN: Brussels, 2005.

[2] Lourenco, P.B., Computations on historic masonry structures. *Progress in Structural Engineering and Materials*, **4(3)**, pp. 301-319, 2002.

[3] Stewart, M.G. & Lawrence, S.J., Model Error, Structural reliability and partial safety factors for structural masonry in compression. *Masonry International*, **20(3)**, pp. 107-116, 2007.

[4] Grubbs, F., Procedures for detecting outlying observations in samples. *Technometrics*, **11(1)**, pp. 1-21, 1969.

[5] Ang, A.H.S. & Tang, W.H., *Probabilistic Concepts in Engineering Emphasis on Applications to Civil and Environmental Engineering*, John Wiley & Sons: USA, 2007.

[6] Stewart, M.G. & Lawrence, S., Structural reliability of masonry walls in flexure. *Masonry International*, **15(2)**, pp. 48-52, 2002.

[7] Holicky, M., Middleton, J. & Vorlicek, M., Statistical analysis of partial safety factors for structural masonry. *Computer Methods in Structural Masonry 4: Proc. of the Fourth Int. Symp. on Computer Methods in Structural Masonry*, eds. G.N. Pande, J. Middleton & B. Kralj, Taylor & Francis: London, pp. 325-338, 1998.

[8] ICOMOS, Recommendations for the analysis, conservation and structural restoration of architectural heritage, International council on monuments and sites: Paris, pp. 37, 2003.

[9] Holicky, M., Pume, D. & Vorlicek, M., Masonry strength determination from tests. *Computer Methods in Structural Masonry 3 – Proc. of the Third Int. Symp. on Computer Methods in Structural Masonry*, eds. G.N. Pande & J. Middleton, Books & Journals International: Swansea, pp. 107-116, 1997.

[10] EN 1990, EN 1990:2002 Eurocode – Basis of structural design, CEN: Brussels, pp. 87, 2002.

Analysis of historic mortars from the archaeological site of Logos and design of repair materials

M. Stefanidou, V. Pachta & I. Papayianni
Aristotle University, Thessaloniki, Greece

Abstract

The ancient city of Logos was one of the most important cities of Macedonia. During the 4th century BC, the settlement formed a city with acropolis, fortifications and a central paved street, alongside of which houses and storages were built. The aim of the study was to analyze mortar samples taken from two houses of the site from the 4th century AD and propose repair materials compatible to the old ones. A series of laboratory tests were carried out in order to determine their physico-mechanical and chemical properties, such as microstructure, porosity, apparent specific gravity, aggregates gradation, compressive strength as well as chemical composition. From the evaluation of the results, comparative observations among samples were made, regarding their structure and constituents, as well as proposals for repair materials. The design of repair materials was based on the principle of compatibility, by retaining their constituents and proportions in the mixtures, while the specific environmental conditions of the site were also taken into account.
Keywords: Logos, historic mortars, repair materials, compatibility.

1 Introduction

Mortars constitute a significant building material, used in constructions even from prehistoric times [1–3]. Their structure and properties usually differ according to their functional role in constructions (structural mortars, renders-plasters, flooring), the technology of each era and the availability of raw materials [1, 3]. Their study reveals a great source of information regarding their structure (type and proportions of raw materials), their properties, as well as their application technology [1–12].

A holistic methodology has been developed in the Laboratory of Building Materials of the Aristotle University of Thessaloniki during the last twenty years and has been applied to the analysis of more than 2000 mortar samples of approximately 300 monuments and historic buildings in Greece [1, 3, 13–17]. The methodology comprises the analysis of the microstructural, physico-mechanical and chemical characteristics of mortar samples, the evaluation of results and the design and laboratory production of compatible repair mortars by retaining the physico-mechanical and chemical properties of the old mortars.

The paper focuses on the application of the methodology to mortar samples taken from archeological site of Logos. The archeological site of Logos is located in central Macedonia, Greece, near the modern town of Edessa. It constitutes one of the most ancient and important cities of Macedonia, since it was a passage connecting the flat with the highlands of Macedonia [18, 19]. The earliest habitation of the site is dated during the Early Bronze Age (3rd millennium BC). During the 4th century BC, the settlement formed a city with acropolis, fortifications, a central paved street, alongside of which houses and storages were built. The city retained its importance until the Early Christian period (7th C AD), when it was restricted to the area of the contemporary city.

The aim of the study was to analyze seven mortar samples taken from two houses, built alongside the central paved street, in order to propose repair materials compatible to the old ones. The extreme environmental conditions of the site were also taken into account as the temperature is below 0°C very often during the winter and frost cycles take place while, during the rest of the year, the humidity is very high. The main pathology symptoms encountered in the building materials are:

- Absence of joints due to mortars' loosening (fig. 1);
- Intense cracking in bricks and stones due to intense temperatures alterations and frost action (fig. 2);
- Biological growth;
- Color alteration of stones due to black crust deposition.

Figure 1: Absence of joints due to mortars' loosening.

Figure 2: Cracks in stones.

2 Materials and methods

Mortars were carefully selected and sampled, in order to be authentic and not from subsequent repair work (valuable information about their authenticity were given by the responsible scientists of the site). The walls consisted of shaped stones, mainly travertine. The joining mortar was absent in many places as shown in figs. 1 and 2. Pieces of bricks the height of which were 2.5–3 cm were encountered between the stones. In situ macroscopic observation assisted the understanding of the deterioration degree and pathology symptoms of the building materials.

A series of laboratory tests were carried out in order to determine the physico-mechanical and chemical properties of the mortars. Microstructure observations were performed with a stereoscope (Leica Wild M10) assisted by image analysis (ProgRes). Aggregates granulometry was tested by hand grounding and sieving, according to EN1015-1:1998. The porosity measurement was performed according to RILEM CPC 11.3, while the compressive strength was determined by testing shaped cubic samples of $4 \times 4 \times 4$ cm^3. Finally, wet chemical analysis assisted by Atomic Absorption (Perkin Elmer 3110) and HPLC (for detecting soluble salt content) was performed on a fine fraction of the sample (<75μm) [17–23].

From the comparative study of the results, the design of the repair mortars proceeded based on the characteristics of ancient mortars (by retaining their constituents and proportions in the mixtures), while the specific environmental conditions of the site were also taken into account. In order to restrict degradation phenomena, a porous mortar structure was favored since this permits the "breathing" of the materials and also is in agreement with the porous nature of the travertine.

3 Results and discussion

The analysis of the experimental results confirmed two different mortar types. In table 1 the properties of the mortar samples are summarised.

The first category comprised lime-based mortars characterized by a pale brown color and a high proportion of CaO (35–37% w/w). The binder was of mixed type and was characterized as a combination of hydrated lime and clayish material with pozzolanic reactivity (due to the high content of silica and alumina), in a proportion 1:0.5. The aggregates were of siliceous origin, with gradation 0–8 mm and their B/A ratio was 1/2.5. A small percentage of aggregates (estimated at 5% of the whole) was fine crushed brick of gradation 2–4 mm. The compressive strength rated from 1.2 to 1.5 MPa and the porosity was 18–22%.

The second category comprised a clay based mortar of dark brown color. The content in CaO was lower than that of the first category (32-34% w/w) while the silica and alumina content was higher (34-35% w/w). The aggregates were of natural origin and coarse (up to 1 cm), while the B/A ratio was 1/2. The compressive strength was low (< 1 MPa), with a respectively high porosity (23–25%).

Table 1: Physico-mechanical and chemical properties of mortars.

Mortar type	Binding system L= lime S= Soil	Total oxides (%w/w)		Aggregates type/ gradation	Color	Compressive strength (MPa)	Porosity (%)	B/A ratio
		CaO	SiO_2+ Al_2O_3+ Fe_2O_3					
I	L/S=1:0.5	35–37	28–29	Siliceous 0–8mm	pale brown	1.2–1.5	18–22	1 :2.5
II	L/S=0.5:1	32–34	34–35	Siliceous 0–10mm	dark brown	<1	23–25	1:2

4 Design of repair mortars

From the evaluation of the results, it was concluded that all mortars were made from the same type of raw materials (binders, aggregates). However, the binding system, the B/A ratio as well as the aggregates granulometry slightly differed.

The aim of the repair materials' proposal was to fulfill the requirements of compatibility and resistance to the extreme environmental factors, retaining the physico-mechanical and chemical properties of old mortars. The following parameters were taken into account [1, 13–15]:

- binder system;
- granulometry and type of aggregates;
- B/A ratio;
- color and texture;
- mechanical strength and porosity level.

In order to proceed with the design of repair materials, local raw materials (soil from the site and aggregates from the area) were analyzed, to test their suitability for using them in the repair mortars. Tests concerning the granulometry, salt content, chemical composition and color hue of the materials were performed (table 2). The results showed that local raw materials were appropriate for the production of repair mortars. The soil composition was close to that obtained from the analysis of the mortars categorized as type II (table1). The utilization of the local materials also contributed to the cost benefit of the site works.

Table 2: Characteristics of local raw materials.

	Total oxides (%w/w)		Soluble Salts (%w/w)		
	CaO	$SiO_2+Al_2O_3+Fe_2O_3$	Cl^-	NO_3^-	SO_4^{2-}
Soil	32.9	32.9	0.00	0.00	<0.01
Sand	36.4	27.5	0.00	0.01	<0.01

The proposals for repair materials were based on traditional binders, such as hydrated lime, natural pozzolan and clay, as well as natural aggregates of local origin and siliceous context. Natural pozzolan was added in order to contribute to the strength development by the pozzolanic reaction with the lime. The clay was added for the color match; its composition shows that it can also contribute to the pozzolanic reaction. In order to reduce the required water, the use of a sulphate free superplasticizer (1% w/w of binders) was proposed. In the proposal for mortar type II which was a clay- based mortar, a small quantity of white cement was added (20% w/w of binders), in order to enhance the mortar's properties in fresh and hardened state (increase early strength development without altering significantly the physical and structure properties of the mortars).

The proposed composition for repair mortars, according to the mortars' type are presented in table 3.

Table 3: Constituents and proportions of repair mortars.

Mortar type	Raw materials	Parts by weight
I	Hydrated lime (powder)	1.0
	Natural pozzolan	0.6
	Local clay (<0.25mm)	0.4
	Sand of natural origin (0-4mm)	3.35
	Gravel of natural origin (4-8mm)	1.5
	Crushed brick	0.15
	Super-plasticizer 1% w/w of binders	
II	Local clay (<0.25mm)	0.8
	White cement	0.2
	Sand of natural origin (0-4mm)	1.8
	Gravel of natural origin (4-8mm)	0.2
	Super-plasticizer 1% w/w of binders	

The proposed mixtures were laboratory produced and their physico-mechanical properties at the age of 28 days are presented in table 4. The open porosity was kept high, the color was compatible with that of the old mortars and so was the strength level. The aggregate gradation was also similar.

Table 4: Physico-mechanical properties of repair mortars.

Mortar type	Flexural strength (MPa)	Compressive strength (MPa)	Open porosity (%)
I	1.0	2.5	19.5
II	0.7	1.5	24

For the in situ application, technical guidelines were given concerning the mixture of raw materials, the water demand, as well as the curing conditions of fresh mortars. The restored walls were kept wet for seven days in order to avoid rapid drying which could cause early cracking. The proposed mortars were successfully applied and monitored before and after application (fig. 3).

Figure 3: Historic masonry before (left) and after (right) restoration.

5 Conclusions

In the case of Logos, scientific knowledge was successfully combined with site experience for the benefit of the heritage structure. In that way, conservation works can be of low cost as local materials can be upgraded and also they can give aesthetically good results. Keeping in mind that ancient builders were searching in the environment of the structure for raw materials, it seems that the same principle can still be effectively applicable. It is important to perform preliminary laboratory tests on the raw materials in order to avoid problems in the structure. After the end of the conservation works, the site should be monitored in order to check the durability of the materials and the affect of the atmosphere on the repair materials.

References

[1] Papayianni, I., The longevity of old mortars. *Applied Physics A, Materials Science & Processing*; **83**, pp. 685-688, 2006.

[2] Van Hees, R.P.J., Binda, L., Papayianni, I. & Toumbakari, E., Characterisation and damage analysis of old mortars. *Materials and Structures*, **37**, pp. 644-648, 2004.

[3] Pachta, V., Study of the technological evolution of mortars. PhD Thesis, Dept. of Civil Engineering, Aristotle University of Thessaloniki, Thessaloniki, 2011.

[4] Moropoulou, A., Polikreti, K., Bakolas, A. & Michailidis, P., Correlation of physicochemical and mechanical properties of historical mortars and classification by multivariate statistics. *Cement and Concrete Research*, **6**, pp. 891–898, 2003.

[5] Maravelaki-Kalaitzaki, P., Bakolas, A. & Moropoulou, A., Physico-chemical study of Cretan ancient mortars. *Cement and Concrete Research*, **33**, pp. 651-661, 2003.

[6] Bugini, R. & Salvatori, A., Investigation of the characteristics and properties of 'Cocciopesto' from the ancient Roman period. *Conservation of Stone and Other Materials*, **l**, pp. 386-393, 1993.

[7] Middendorf, B., Hughes, J.J., Callebaut, K., Baronio, G. & Papayianni, I., Investigative methods for the characterisation of historic mortars-Part 1: Mineralogical characterisation. RILEM TC 167-COM: 'Characterisation of Old Mortars with Respect to their Repair'. *Materials and Structures*, **38**, pp. 761-769, 2005.

[8] Miriello, D., Barca, D., Bloise, A., Ciarallo, A., Crisci, G.M., De Rose, T., Gattuso, C., Gazineo, F. & La Russa, M.F., Characterisation of archaeological mortars from Pompeii (Campania, Italy) and identification of construction phases by compositional data analysis. *Journal of Archaeological Science*, **37**, pp. 2207-2223, 2010.

[9] Baronio, G., Binda, L., Tedeschi, C. & Tiraboschi, C., Characterisation of the materials used in the construction of the Noto Cathedral. *Construction and Building Materials*, **17**, pp. 557-571, 2003.

[10] Pavía, S. & Caro, S., An investigation of Roman mortar technology through the petrographic analysis of archaeological material. *Construction and Building Materials*, **22**, pp. 1807-1811, 2008.

[11] Middendorf, B., Hughes, J.J., Callebaut, K., Baronio, G. & Papayianni, I., Investigative methods for the characterisation of historic mortars- Part 2: Chemical characterisation. RILEM TC 167-COM: 'Characterisation of Old Mortars with Respect to their Repair'. *Materials and Structures*, **38**, pp. 771-780, 2005.

[12] Baronio, G., Binda, L. & Tedeschi, C., Thick mortar joints in Byzantine Buildings: study of their composition and mechanical behaviour. *Proc. of International Conference on Studies in Ancient Structures*, Istanbul, pp. 235-244, 1997.

[13] Papayianni, I., Pachta, V. & Stefanidou, M., Analysis of ancient mortars and design of compatible repair mortars: the case study of Odeion of the Archaeological site of Dion. *Construction and Building Materials*, **40**, pp. 84-92, 2013.
[14] Papayianni, I., Design of compatible repair materials for the restoration of monuments. *International Journal for Restoration*, **1-6**, pp. 623-636, 2004.
[15] Papayianni, I. & Stefanidou, M., Mortars for intervention in monuments and historical buildings. *STREMAH 2003*, Chalkidiki Greece, ed. C.A. Brebbia, pp. 57-64, 2003.
[16] Papayianni, I., Pachta, V., Stefanidou, M. & Konopissi, V., Technological Evolution of Structural Mortars. *Proc. of 8th Int. Conference SAHC 2012*, Structural Analysis of Historical Constructions, Wrocław, Poland, ed. J. Jasienco, pp. 1248-1254, 2012.
[17] Papayianni, I. & Stefanidou, M., Porosity and structure of old mortars. *Proc. of 2nd Int. Congress on Studies in Ancient Structures*, Istanbul Turkey, eds. G. Arun & N. Seckin, pp. 509-518, 2001.
[18] Greek Ministry of Culture "Ancient Edessa" IZ Euphoria of Pre-historic and Classical Antiquities.
[19] Chrisostomou A. Edessa in the Early Christian age through the archaeological findings of the last decade. AEMΘ 1996; 10 A: 173-187.

Infrared thermograph image analysis for the identification of masonry coatings in historic buildings, in relation to several samples prepared as patterns

L. Palaia, J. Monfort, P. Navarro, R. Sánchez, L. Gil, A. Álvarez, V. López & S. Tormo
Polytechnic University of Valencia, Spain

Abstract

The application of non-destructive methods for architectural heritage knowledge is of extreme importance, in order to avoid producing irreparable damage. The methods to gain this knowledge should be complementary to other types of studies and tests, to make the process of building assessment easier. The detection of previous openings of doors and windows, walls that have been demolished and later disguised under mortar coatings, allows interpretation to act on those points if necessary, without affecting the conservation of the original coatings existing in the building. It is possible to achieve this aim using infrared thermography (IRT); this research group has applied it successfully in some historic buildings. Some samples have been prepared in the laboratory, consisting of bricks coated with different types of mortars hiding iron plates. Later, IRT images of these samples were captured, which were compared with IRT images of the facades of an existing building. Conclusions were reached from this experience that could be extended to other situations.
Keywords: infrared thermography, façade coatings analysis, non-destructive diagnosis, historic buildings.

1 Introduction

The evaluation of historic buildings on site using destructive techniques is against the proper attitude for achieving conservation aims. The development of non-destructive evaluation (NDE) techniques should be considered essential.

Infrared thermography (IRT) is applied for non-destructive evaluation in many fields. One of these is the surveying of ancient buildings [1–3], producing no damage to the existing construction elements. Research carried out in this field shows that this method could be used as a non-destructive approach to building surveying [4, 5].

According to Maldague [6], each NDE technique has its own strengths and weaknesses. In the case of thermography, the strengths are a fast inspection rate, no contact being established with the building inspected; safety, as no harmful radiation is involved, and results relatively easy to interpret (image format). However, it is also important to consider the effects of thermal losses (due to convection, radiation and conduction), perturbing thermal contrasts; the cost of the equipment (basically of the IRT camera) and that this technique enables the detection of only subsurface defects resulting in a measurable change of thermal properties.

As mentioned above, the technique involves the measurement of radiated electromagnetic energy emitted by a surface at its temperature. It is called the spectral radiance and is governed by Planck's law. Infrared thermographic images are produced by the emitted radiation E of a solid at a specified temperature, per unit surface and time; this value is given by the Stefan-Boltzman expression [7]

$$E(T) = \varepsilon(T)\sigma T^4 \tag{1}$$

where T is the absolute temperature of the solid, σ is the Stefan- Boltzman constant, independent of the solid material and its temperature T, and ε is the material's emissivity. Its value is between 0 and 1, being 1 when it describes a solid with the maximum emissive capacity, the so-called black body.

Emissivity also depends on temperature, though, within specified temperature ranges, it may be considered as a constant for the great majority of solids. This is the case of the building materials commonly used in construction. So, for example, it may be accepted that the emissivity of building materials does not vary significantly within the temperature range corresponding to ambient variations. It also may be assumed that the emissivity of the common building materials lies between 0.7 and 0.9 (except for metallic materials, whose emissivity is lower than these values).

When analysing an IRT image for the assessment of an architectural surface, two factors should be considered: the temperature and the emissivity. For that reason, a qualitative IRT analysis, based on the intensity of the images produced by differences between radiated emissions by architectonic elements situated close to the surface, must be performed considering that the variations shown in the image may be produced by a difference in the emissivity or by a difference in temperature. Obviously, these differences could only be detected between building elements of different nature existing in the same surface.

The difference in temperature between a building element and any point adjacent to it will depend, basically, on its mass, its specific heat or thermal conductivity, properties that may produce a substantial variation in its thermal inertia with respect to its adjacent points. This different thermal inertia will produce variable differences of temperature between this element and its surrounding area, in the heating and cooling processes of the building façade. For this reason, a façade assessment requires the analysis of several IRT images obtained at different moments of the heating or cooling processes.

2 Historic buildings survey

Historic buildings should be preserved from destruction. They need to be conserved, maintained, repaired or restored. This means that no damage must be introduced to the buildings during their assessment. On many occasions, some exploration work is done to gather information about the building materials, the previous layout and structural organization. It is useful to reduce this exploratory work in a way that would lead to less destruction of the original elements. IRT images can be used for this purpose.

3 Previous experiences

The first approaches with the IR thermocamera gave interesting results for the detection of building material discontinuity under the façade coatings. This device was able to capture images of the building walls, producing thermal images showing lintel elements as in figs. 1 and 2, also previous repairs and signs of demolished walls on the existing ones, as in figs. 3 and 4. A previous window that had been walled up was also identified, as in figs. 4–6.

Figure 1: Existing window.

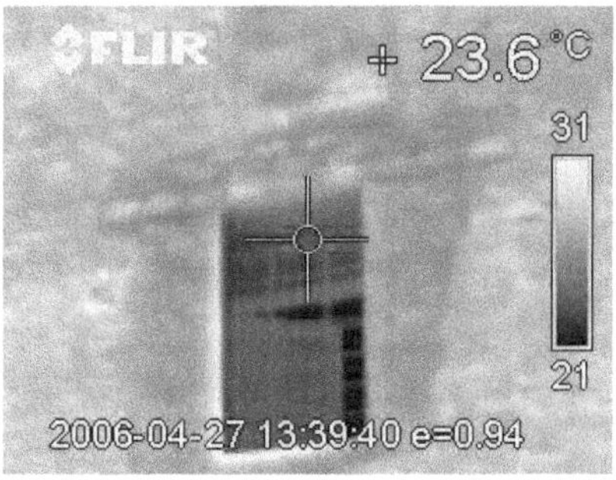

Figure 2: IRT image showing the lintel.

A medieval tower was surveyed by means of IRT images, with promising results, so the research group resolved to improve the applied method. In this tower, located at Torrent village, close to Valencia City [7], different IRT images were captured in the same day, at different hours, in order to detect the best time to capture IRT images. The factors considered for the IRT images analysis were insulation level, orientation and wind velocity.

Images were compared and it was determined that the best hour of the day to take the IRT images was from 20:00 hours onwards, in summer time, when the building materials were cooling down after a sunny day. It was possible to identify walled up hollows and to determine the humidity presence.

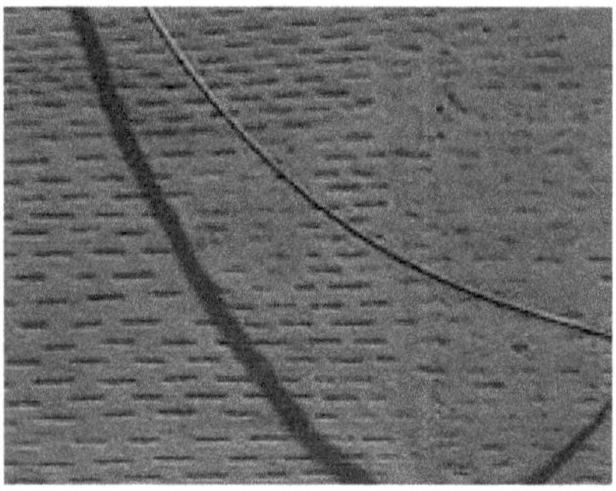

Figure 3: Image of the wall.

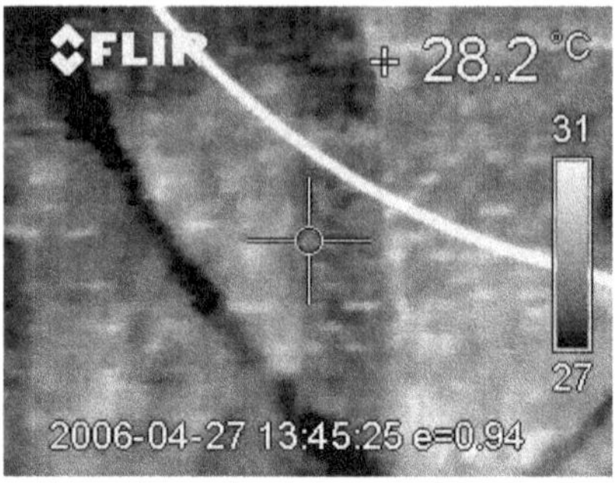

Figure 4: IRT image showing a previous transversal wall, today demolished.

Figure 5: Image of a wall.

Figure 6: IRT image showing a walled up window.

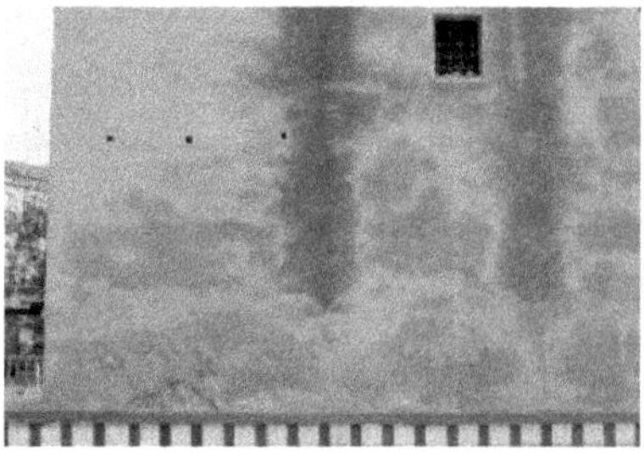

Figure 7: Detail of the tower.

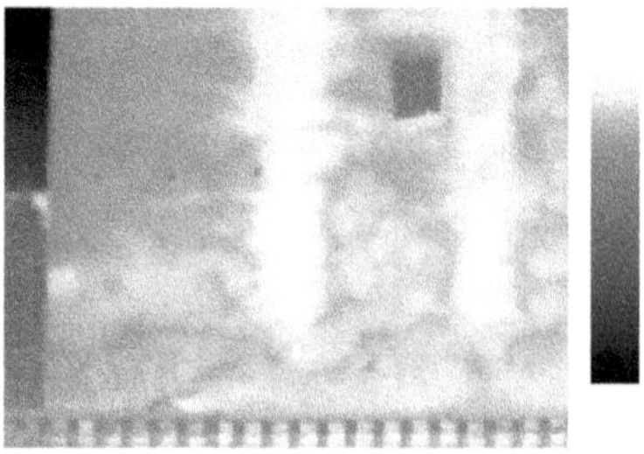

Figure 8: IRT image of the same area, showing a previous opening.

4 Experimental procedures

As mentioned earlier, the majority of building materials usually employed in construction have emissive values, lying within the 0.7–0.9 range. Different mortar coatings have special relevance because of their frequent use in historic building facades.

An IRT analysis of the mortar coatings of historic buildings demands a preliminary study of the emissive behaviour of different mortar types. For this purpose, a comparative study of several mortar types was carried out as part of this research, in order to quantify the thermographic appraisal of their possible emissive differences.

The passive infrared thermographic approach was employed using a B2 thermocamera by FLIR System, with field of view/min focus distance of 34°×25°/0.1 m Automatic, thermal sensitivity <0.10°C at 25°C, a Detector Type Focal plane array (FPA) uncooled micro bolometer and spectral range 7.5 to 13 μm. The temperature range of the IRT detector extends from –40 to +300°C.

Mortar samples were prepared with the following characteristics: manual bricks, measuring: 22.5×10.8×3.4 cm^3; wooden frames to provide uniform thickness of mortar coatings, 4.6, 5.5×6.5×3 cm^3 (section); iron plates, 30×120×2 mm^3; current sand for mortar coatings, cement CEM II-B L 32.5 N; gypsum YG and soaked lime.

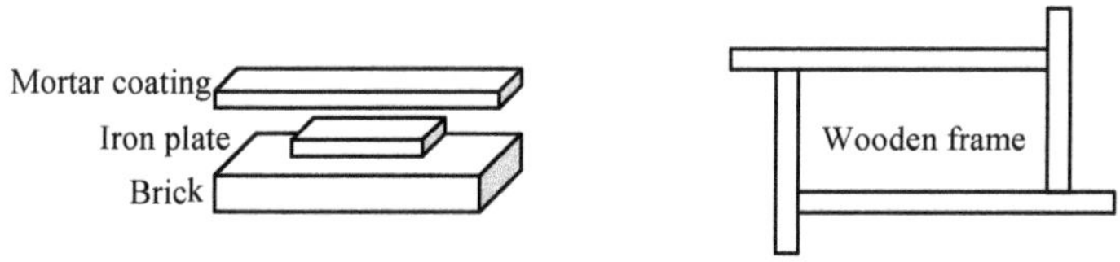

Figure 9: Samples scheme.

The mortars prepared as samples were: cement mortar, 1:3; lime mortar, 1:3; gypsum mortar, 1:2, with increasing thickness as shown in table 1. Iron plates were introduced beneath the mortar coatings in nine of the mortar samples (1 to 9), with the aim of detecting their presence thermographically. The preparation of the samples can be seen in figs. 10–15. The samples exposure to sun radiation is shown in fig. 16 and the corresponding IRT images at 16:30 hours are shown in figs. 17 and 18.

Table 1: Samples characteristics.

Sample N°	Mortar type	Thickness	Iron plate
1, 4, 7	Cement mortar	1, 2 and 3 cm	yes
10	Cement mortar	3 cm	no
2, 5, 8	Lime mortar	1, 2 and 3 cm	yes
11	Lime mortar	3 cm	no
3, 6, 9	Gypsum mortar	1, 2 and 3 cm	yes
12	Gypsum mortar	3 cm	no

Samples 1, 4, 7 and 10 were prepared with cement mortar, those numbered 2, 5, 8 and 11 were prepared with lime mortar, and finally, samples 3, 6, 9 and 12 were prepared with gypsum mortar, always varying the thickness: 1, 2 and 3 cm. Samples 10, 11 and 12 did not contain iron plates and have a 3 cm thick coating. IRT images were captured at different times: 15:30, 16:30 and 17:30. The obtained results were very revealing.

Figure 10: Wooden frame preparation.

Figure 11: General view of the samples with the bricks and iron plates present.

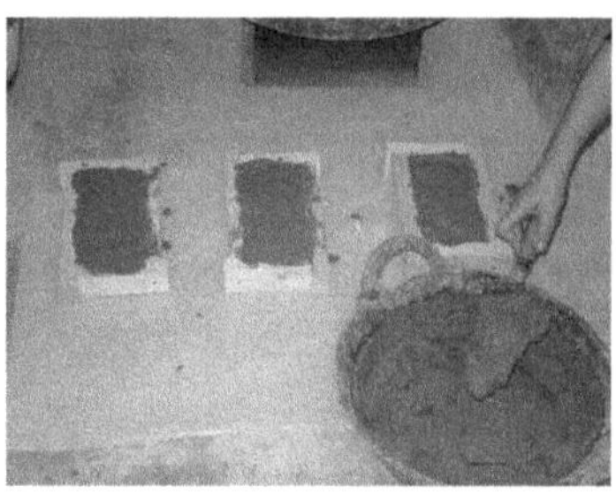

Figure 12: Cement mortar samples.

Figure 13: Lime mortar samples.

Figure 14: Gypsum mortar preparation.

Figure 15: Gypsum mortar samples.

Firstly it should be mentioned that the temperature reached by the samples of the same thickness was the same. For this reason, the different emissivity must be analysed between samples of the same thickness. This can be seen in the first group of samples, numbered 1 to 3; the sample with the highest emissivity is the cement mortar type (n° 1) and the sample with the least emissivity is the gypsum mortar type (n° 3). The same behaviour is clearly observed in samples n° 4–6, where the thickness of the mortar coating is 2 cm. For samples n° 7–9 and 10–12, the differences between emissivity are not detectable. Considering that all of these six samples were coated with 3 cm mortar thickness, their temperature is less than that of the first six samples, n° 1–6, with 1 and 2 cm coating thickness.

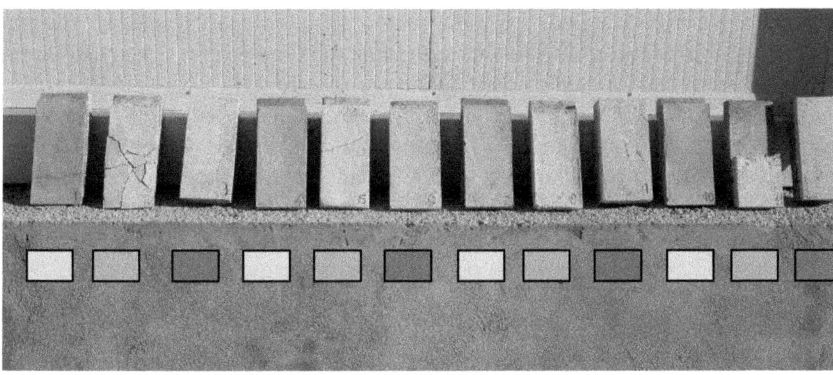

Figure 16: General view of the mortar samples.

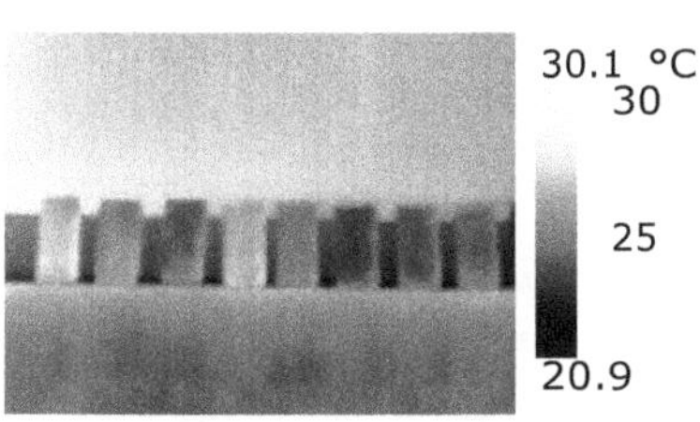

Figure 17: IRT images, samples 1–8.

Figure 18: IRT images, samples 6–12.

At this point, a very important conclusion can be reached: the differences between the emissivity of different coatings of various compositions at the same temperature are clearly identified in an IRT image at a high enough temperature. Cement mortar coatings present higher emissivity than lime mortar coatings and the latter show higher emissivity than those made with gypsum mortar. A clear value where this step is produced cannot be determined.

However, it can be established as a second conclusion that the IRT image of a building façade will be significant if it has reached a minimum temperature with enough uniformity. For this reason, the IRT image should be obtained after the heating process of the surface, and, as mentioned previously, in different moments of the cooling down process to evaluate the different thermal inertia effect of its elements.

It was not possible to identify the iron plates through the thermographic images. It must be considered that the emissive values are measured under conditions of emissivity involving radiation in the open air. Therefore, even with the metallic element emissivity being lower to that of the mortar coating it, the less radiation of the iron plate is masked by the emissivity of the mortar covering. This issue should be examined further, but an initial hypothesis about it may be established, by proposing that hidden elements could be detected when they have enough mass with a different specific heat or thermal conductivity, to show a different temperature to their coatings, so a different emissivity from that emitted by their coatings could be seen.

A better approach whereby this behaviour could be detected is when the mortar coatings are applied to a historic building façade. In this way, the adjustment to the heating process of the façade is simulated with one of the mortar samples, avoiding the isolated heating effects of the samples previously considered.

The next phase of the research was to apply the same mortar coverings over the existing walls of an ancient building, located at Picanya (fig. 19), which is going to be reused as an administrative centre. It is a free standing building, with orientation to all directions in the four façades and different construction systems in the walls, such as brick walls, pisè, and masonry work.

As the wall coatings had deteriorated, it was possible to prepare "in situ" samples, 30×30 cm^2, 1 cm thick, using the same proportions as of those prepared in the laboratory. A small coin was introduced at the centre of each sample, in order to detect them with the IRT images.

Figure 19: Picanya – historic building. Principal façade (east).

Figure 20: Mortar samples at the south façade.

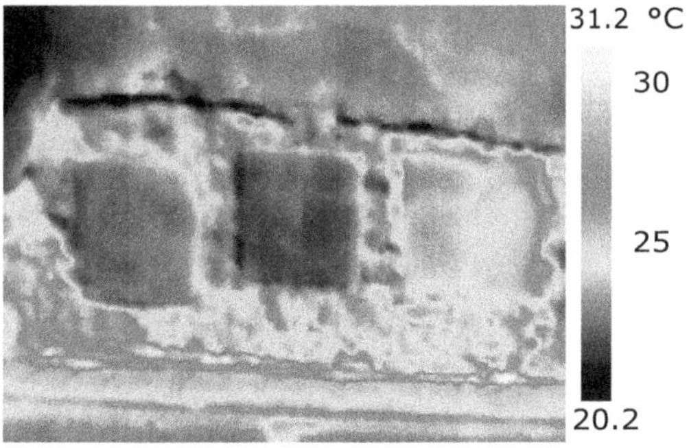

Figure 21: IRT image of the mortar samples at the south façade.

Figure 22: Samples at the west façade.

The emissive behaviour of the different mortar types can be observed in these ITR images of the samples as before, so that the conclusions reached previously can be valid for mortar coatings applied to the walls. When the building materials present a high emissivity in the same temperature ranges, the different composition can be clearly detected by IRT images.

This work opens an interesting line of research in IRT images of historic buildings analysis based on the comparison of IRT images with qualitative sample images of materials or elements previously studied.

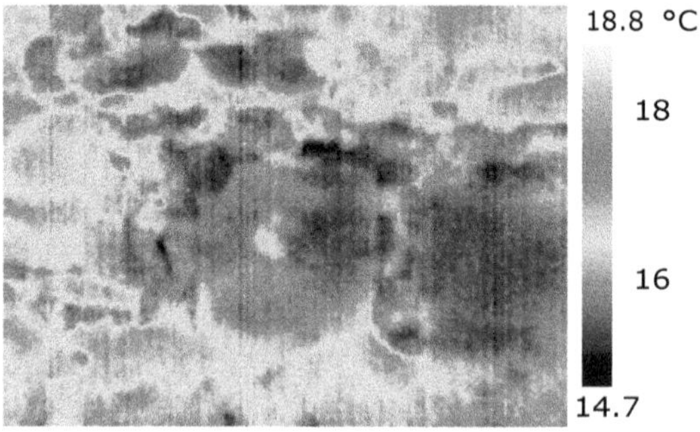

Figure 23: IRT image of the mortar samples at the west façade.

Acknowledgements

The authors are professors at the Polytechnic University of Valencia and members of the I+D+i Project financed by the Ministerio de Fomento, 2004 (Spain).

References

[1] Rosina, E., Using infrared thermography to detect moisture in historic masonry: a case study in Indiana. *APT Bulletin*, **34(1)**, pp. 11-16, 2003.
[2] Sandrolini, F. & Franzoni, E., An operative protocol for reliable measurements in porous materials of ancient buildings. *Building and Environment*, **41(10)**, pp. 1372–1380, 2006.
[3] Schellen, H.L. & de Wit, M.H., Heat and moisture modelling of a monumental, massive building. *5th Symposium on Building Physics in the Nordic Countries*, Gothenburg, Sweden, 1999.
[4] Moropoulou, A., Avdelidis, N.P., Delegou, E.T. & Koui, M., Infrared thermography in the evaluation of cleaning interventions on architectural surfaces. *Proceedings Inframation 2001*, Infrared Training Centre, pp. 171–175, 2001.
[5] Moropoulou, A., Delegou, E.T., Avdelidis, N.P. & Koui, M., Assessment of cleaning conservation interventions on architectural surfaces using an integrated methodology. *Materials Issues in Art and Archaeology VI*, eds. P. Vandiveer, M. Goodway, J.R. Druzik & J.L. Mass, Materials Research Society Publications, USA, pp. 69–76, 2002.
[6] Maldague, X.P., *Theory and practice of infrared technology for nondestructive testing*, Wiley: New York, 2001.
[7] Maldague, X.P., *Nondestructive evaluation of materials by infrared thermography*, Springer-Verlag: London, 1993.

[8] Palaia, L., Sánchez, R., Monfort, J., Gil, L., Álvarez, Á., Tormo, S., López, V., Pérez, C. & Navarro, P., Aplicación de la termografía infrarroja a los estudios previos de los edificios históricos, para la identificación de materiales y elementos constructivos ocultos tras los revestimientos, 11° Congreso Español de NDE, Gijón,. 2007 (submitted).

Shear seismic capacity of tuff masonry panels in heritage constructions

G. Marcari[1], G. Fabbrocino[1] & G. Manfredi[2]
[1]*SAVA Department, University of Molise, Termoli, Italy*
[2]*DIST Department, University of Naples, Naples, Italy*

Abstract

Tuff masonry structures have been built since old times in countries located in the Mediterranean areas; they represent a significant part of the existing masonry building inventory of Central-South Italy, including historical architecture. Due to a lack of knowledge in relevant strength and deformability parameters for tuff masonry, experimental and numerical analyses concerning their shear response are certainly of interest. The present paper focuses on single and multiple-leaf tuff masonry walls under various in-plane loading conditions. Experimental displacement-controlled results on large specimens have been used to calibrate finite element (FE) numerical models. A macro-modelling approach is used, which is specifically based on a composite plasticity model under plane stress conditions. Comparisons between numerical and experimental results are provided. The ability of the proposed model to fit the overall performances of tuff panels is thus demonstrated. Additional and relevant information about the relation between mechanical parameters of tuff masonry and the corresponding seismic capacity are given for safety assessment and retrofit design purposes.
Keywords: masonry, tuff, shear strength, seismic capacity, non linear analysis.

1 Introduction

Masonry structures built since ancient times are generally characterised by high levels of seismic vulnerability, thus they generate significant interest especially when development of analysis, assessment and reliable tools for their seismic protection are considered. The problem is complex for all structural types, but also critical for tuff masonry constructions located in Central-Southern Italy regions and particularly in Molise [1]. The variety of these buildings, most of which were erected according to traditional rules of practice and rarely engineered, requires a rational approach to seismic assessment based on numerical analyses, well supported and validated by experiments. In recent years numerical strategies and constitutive models for masonry have been developed for the interpretation of its complex experimental behaviour, but a more in-depth knowledge of material behaviour is often required via load-displacement controlled tests. In particular, availability of seismic capacity data is crucial for practical applications and safety evaluations issued by modern performance based codes, that is, US FEMA 356 [2], Eurocode 8 [3] and Italian OPCM 3274 [4]. In addition, validated mechanical parameters and practice-oriented numerical models should be made available on a large-scale for practical applications. Therefore, the present work aims at numerical analysis of the monotonic response of typical ancient tuff masonry walls. In particular, attention is paid to the response of single and multiple-leaf masonry walls under various in-plane loading conditions. Displacement-controlled experiments on large-scale masonry tests, small masonry samples and basic components [5–7] provided an extensive database of properties including strength, stiffness characterization and post-peak response, which have been implemented in finite-element analyses. A macro-modelling approach was followed and the composite plasticity model with Rankine type-Hill type yield criteria for plane stress conditions developed by Lourenço *et al.* [8, 9] was adopted.

Comparisons between numerical results and experimental data point out the ability of the proposed model to fit the overall shear performances of tuff panels. Besides, information about correlations between relevant mechanical parameters and seismic capacity of tuff walls is provided.

2 Finite element analysis

In this section, a numerical investigation of the in-plane response of masonry shear-walls, performed via a continuum based approach, which takes into account the anisotropic behaviour of the masonry, is briefly reported. The DIANA FEM code, version 8.1 [10] has been used. The anisotropic composite plasticity model for plane stress structures introduced by Lourenço *et al.* [8, 9], is adopted. The model is able to reproduce elastic and inelastic behaviour in two orthogonal directions, the orientation of the bed and head joints of masonry. Two failure mechanisms can be taken into consideration: the first associated with localised tensile fracture processes and the second associated with a more distributed fracture process which can be related to crushing of the material. Orthotropic elasticity is combined with orthotropic plasticity. The model includes the Rankine type-Hill type yield criteria formulated in terms of the stress state components with respect to the material axes. The axes of orthotropy *x*-*y* are coincident with the material axes, horizontal (bed) joints and vertical (head) joints, respectively. The inelastic material parameters of the model are the following: f_{tx} and f_{ty} are the tensile strengths, G_{ftx} and G_{fty} are the tensile fracture energies, f_{cx} and f_{cy} are the compressive strengths, G_{fcx} and G_{fcy} are the compressive fracture energies, α is a parameter depending on the contribution of the shear stress to tensile failure, β is a parameter related to the coupling of normal stresses as compressive failure is concerned and γ is a parameter which controls the contribution of the shear stress to compressive failure. The fracture-energy based regularization is adopted to mitigate the sensitivity of the results with respect to the mesh size. To this end, an equivalent length h_c is incorporated in the model depending on the chosen element type, element size, element shape, integration scheme and even on the particular problem considered. In DIANA h_c is related to the area of an element A_e by

$$h_c = \sqrt{2A_e} \, ;$$

besides, for quadratic elements, $\alpha_h = 1$. In the case of multiple-leaf walls, a regular mesh of 30×30 four-node quadrilater isoparametric plane stress element, with a 2×2 Gauss integration scheme was used. The analysis was carried out with the Linear Stiffness iteration method, and the linear stiffness matrix is used all the time. A line search algorithm was used in order to stabilize the convergence behaviour, and it was combined with the arc-length method in such a way that snap-through or snap-backs behaviour can be followed. The control of the numerical process took place by means of load control mode. Boundary supports were given at the base of the multiple-leaf panel such that points were fully pinned. Besides, a rigid connection at the top of the wall leads the joints to exhibit the same horizontal displacement. The loading consisted of a vertical load q uniformly distributed along the upper edge of the panel. Self-weight of the panel was considered first. Later on, a horizontal load was applied on the top of the panel. It is worth noting that the main objective of the present study is not the exact reproduction of a single experimental result, but the calibration of relevant mechanical parameters of typical tuff masonry panels so that the overall shear response can be well reproduced. Therefore, the ability of the model to fit the experimental results is checked independently of uncertainties related to some numerical parameters associated to anisotropy of masonry. In addition, relevant aspects of the mesh dependency of numerical results are also investigated. It is known, in fact, that numerical analyses of strain softening materials can be mesh dependent even if an energy-based regularization is adopted. Measured elastic and inelastic properties have been used. In particular, the elastic modulus E_y and the Poisson ratio ν_{xy} were set equal to the experimental values: $E_y = 630$ MPa, $\nu_{xy} = 0.2$; the calibrated shear modulus $G_{xy} = 70$MPa was assumed ($G_{xy} = E_y/9$), which was close to $E_y/6$ as suggested by relevant codes. The elastic stiffness $E_x = 600$MPa was lower than E_y, according to literature suggestions [11]. The vertical compressive strength was equal to the experimental value $f_{cy} = 1.1$ MPa evaluated over the gross section area, according to relevant codes. This is a rational approach, even in the present case where a full knowledge of the sub-assemblage can give additional information. The compressive strength parallel to the bed joints was lower than f_{cy}, according to technical literature [12]. In particular f_{cx} was set to about $0.5f_{cy}$ ($f_{cx} = 0.5$MPa). The inelastic behaviour under compression of masonry panels was described by the compressive energy fracture G_{fc}. A mean value of $G_{fcy} = 10$ Nmm/mm^2 resulted from experimental stress-displacement relationships given in Marcari *et al.* [5] and Marcari [7].

Such a value is in agreement with those calculated from compressive stress-strain relationship reported in the literature for multiple-leaf tuff panel [13, 14]. Since values of G_{fcx} were not directly measured, G_{fcx} was calibrated and resulted equal to $0.3G_{fcy}$ (G_{fcx} = 2.7 Nmm/mm^2). The masonry tensile strength f_{tx} was then determined as the lowest value corresponding to tension crack in the brick or shear slip along a longitudinal joint, after Schubert and Weschke [15]. From available data on tensile strength of mortar samples and stones, and from triplet test results [6], the calculated tensile strength was f_{tx} = 0.17 MPa. The tensile strength f_{ty} = 0.06 MPa was set close to the empirical value $f_{ty} = (2/3)f_{t,\text{mor}}$ given in Tassios [16], with $f_{t,\text{mor}}$ representing the tensile strength of the mortar. Therefore the following strength ratios were adopted: f_{cx}/f_{cy} = 0.5; f_{tx}/f_{ty} = 2.8, $f_{cy}/f_{ty} \approx 18$, in analogy with similar relations found in masonry technical literature [17] and $f_{cx}/f_{tx} \approx 3$. Other inelastic parameters were not available and were assumed to vary within a physical range. In such a way, a calibration of the values for tuff masonry was obtained. A summary of masonry properties and relevant strength and deformation parameters are given in table 1.

For the numerical analysis of solid walls, a regular mesh of 16×16 four-node quadrilater isoparametric plane stress element, with a 2×2 Gauss integration scheme was used. The equations related to the finite element discretization were solved using an incremental-iterative Linear Stiffness method, with arc-length control and line search technique.

The simulated loading scheme consisted of a vertical compressive force at the upper edge and of the self weight. Proper boundary conditions at the panel edges were considered in order to simulate the effect of the two steel loading shoes.

Table 1: Summary of masonry parameters for non-linear analysis.

Type of masonry panel	Elastic properties	Inelastic properties Rankine type Criterion	Inelastic properties Hill type Criterion	Strength and deformation ratios
Multiple-leaf wall panel	E_x=600MPa E_y=630MPa ν_{xy}=0.2 G_{xy}=70MPa	f_{tx}=0.17MPa f_{ty}=0.06Mpa α_t=1.6 α_h=1.0 G_{ftx}=0.008Nmm/mm^2 G_{fty}=0.09Nmm/mm^2	f_{cx}=0.5MPa f_{cy}=1.1MPa γ=1.2 β=-1.5 G_{fcx}=2.7Nmm/mm^2 G_{fcy}=10Nmm/mm^2	E_x/E_y=0.95 f_{cx}/f_{cy}=0.45 f_{tx}/f_{ty}=2.8 f_{cy}/f_{ty}=18 f_{cx}/f_{tx}=3 G_{fcx}/G_{fcy}=0.27 G_{ftx}/G_{fty}=0.08
Solid wall	E_x=647MPa E_y=680MPa ν_{xy}=0.15 G_{xy}=110MPa	f_{tx}=0.35MPa f_{ty}=0.12Mpa α_t=1.6 α_h=1.0 G_{ftx}=0.008Nmm/mm^2 G_{fty}=0.09Nmm/mm^2	f_{cx}=0.69MPa f_{cy}=2.3MPa γ=1.2 β=-0.6 G_{fcx}=1.6Nmm/mm^2 G_{fcy}=6Nmm/mm^2	E_x/E_y=0.95 f_{cx}/f_{cy}=0.3 f_{tx}/f_{ty}=2.9 f_{cy}/f_{ty}=19 f_{cx}/f_{tx}=2 G_{fcx}/G_{fcy}=0.27 G_{ftx}/G_{fty}=0.08

The elastic modulus E_y and the Poisson ratio ν_{xy} were set equal to the experimental values: E_y = 680 MPa, ν_{xy} = 0.15. The shear modulus was set equal to $E_y/6$ = 110 MPa [4], while the elastic stiffness in the direction parallel to the bed joints E_x was equal to $0.95E_y$ = 647 MPa, in analogy with the FEM analysis of multiple-leaf panels. The experimental value of the vertical compressive strength was used, f_{cy} = 2.3MPa, whereas the horizontal compressive strength had to be calibrated at $0.30f_{cy}$ = 0.69 MPa, so that the experimental behaviour could be correctly reproduced. The compressive fracture energy G_{fcy} of 6 Nmm/mm^2 resulted from the integral of the experimental stress-displacement relationship given in Prota *et al.* [6]. The compressive fracture energy G_{fcx} was about $0.3G_{fcy}$, according to the FEM analysis of multiple-leaf panels (G_{fcx} = 1.6 Nmm/mm^2). Due to the lack of available data on uniaxial tensile behaviour, the tensile fracture energies were considered equal to those of the multiple-leaf specimens, thus, assuming G_{ftx} = 0.008 Nmm/mm^2 and G_{fty} = 0.09 Nmm/mm^2. With reference to the strength ratio f_{cy}/f_{ty} of multiple-leaf panels, the masonry tensile strength f_{ty} has been approximately determined according to $f_{cy}/18$ (f_{ty} = 0.12 MPa). Besides $f_{tx} \approx 2.9f_{ty}$ = 0.35 MPa. Other inelastic parameters were investigated by a numerical study where each parameter was varied in a physically realistic range, and the values corresponding to predictions that best fit the experimental response were selected. Masonry parameters and relevant strength and deformation ratios are given in table 1. The sensitivity of the numerical responses with respect to the uncertain parameters (difficult to be measured in practice) has also been assessed, but not reported herein for the sake of brevity.

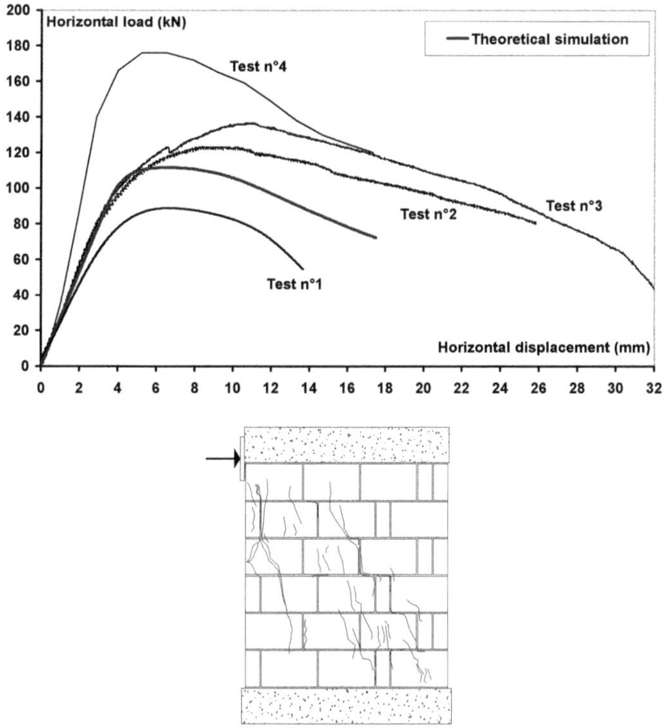

Figure 1: Experimental vs. numerical load-displacement curves and typical crack pattern at failure for multiple-leaf panel.

3 Numerical results

Fig. 1 shows the comparison between the experimental load-displacement curves denoted as test n° 1, 2, 3, 4 and the numerical results. The experimental response is very well reproduced in the elastic range, up to a horizontal displacement δ_x of about 4 mm that basically corresponds to the diagonal shear crack openings of the panels. A good agreement is also found with respect to the collapse load, which is found about 15% lower than the mean experimental value.

An overall good agreement is found even in the post-peak phase. The numerical response of the wall is depicted in fig. 2, in terms of deformed meshes and principal stresses at displacements of 4 and 14 mm, corresponding to 0.25% and 0.89% drift δ_x/H ratios, respectively. During the entire loading process, cracks in the panels developed in a large band that goes through the top-left and the bottom right corner. This again agrees with the general trend of experimental results. It was observed that a large reserve of strength was shown and a concentration of the minimum principal stresses in the shear band, that lead to local crushing of the corners. Therefore, the failure mechanisms observed in the experiments are well captured by the model.

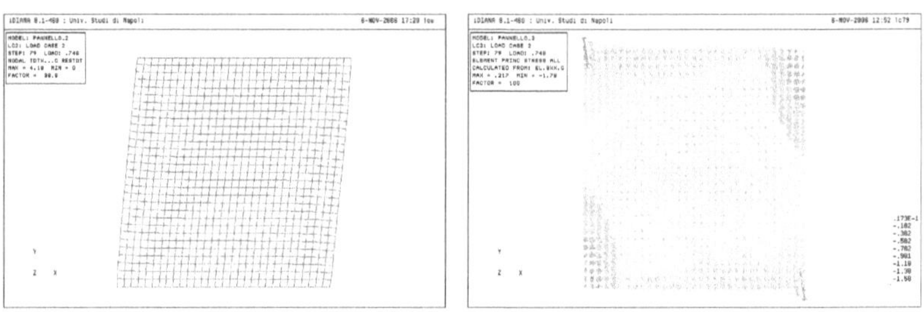

δ_x=4 mm – δ_x/H=0.25%.

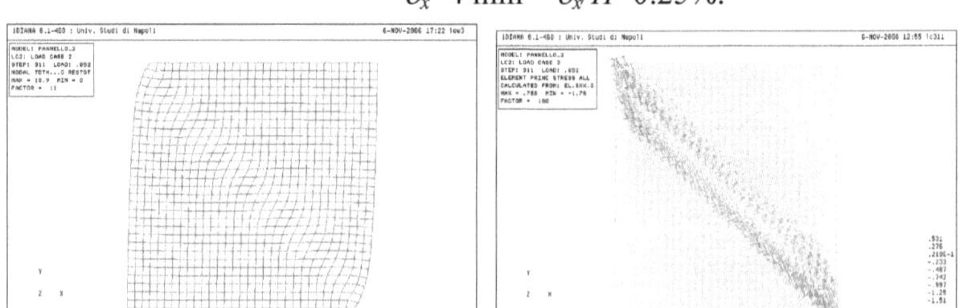

δ_x=14 mm – δ_x/H=0.89%.

Figure 2: Deformed mesh and principal directions of stresses at different drifts $\delta x/H$.

Fig. 3 shows the comparison between the experimental shear stress versus vertical deformation and the numerical results of solid wall panels. The experimental curves are denoted as test n° 1, 2, 3, 4. The predicted shear strength was computed as $\tau = 0.707V/A$, where V is the applied load and A the cross sectional area ($A = 0.257\ m^2$), according to ASTM E51981 [19].

Comparing results in terms of initial stiffness, it is possible to observe that the numerical response well reproduces the elastic behaviour, up to about ε_v =-0.005 that basically corresponds to the first cracking load. Another important aspect is that a good agreement was achieved in terms of strength, since the numerical peak strength is found only 12% higher than the mean experimental value.

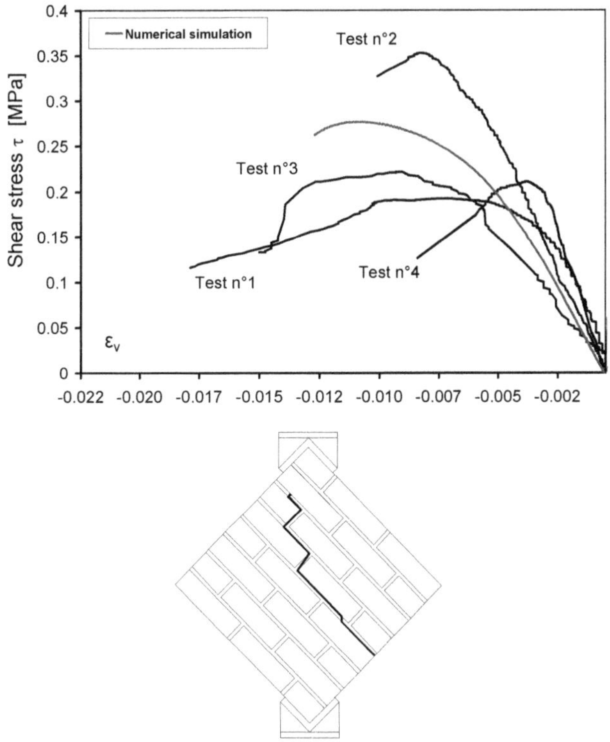

Figure 3: Experimental vs. numerical load-displacement curves and typical crack pattern at failure for solid masonry panel.

Again, the softening behaviour is well captured by numerical analyses in analogy with results obtained for multi-leaf panels. The behaviour of the solid wall is depicted in fig. 4 in terms of deformed meshes and principal stresses at vertical displacements of –1.43 mm and –4.9 mm, corresponding to the cracking load and the ultimate load, respectively.

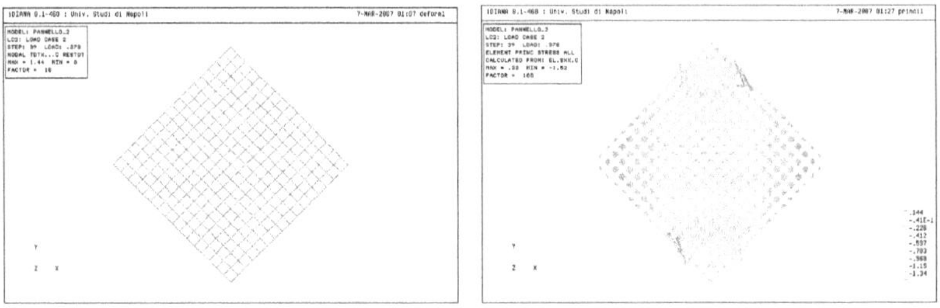

δ_y =-1.43 mm – cracking load.

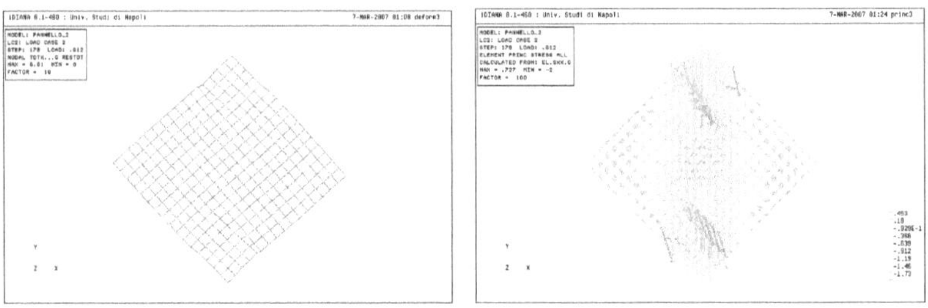

δ_y =-4.9 mm – ultimate load.

Figure 4: Deformed mesh and principal directions of stresses at different vertical displacement δy.

(a) (b)

Figure 5: Mesh sensitivity of the numerical results for multiple-leaf wall (a); solid wall (b).

The crack pattern obtained by macro-modelling mainly arises in a large band that goes through the top and the bottom corner, combined with crushing cracks occurring near the steel loading shoes. The difference found between predicted and observed failure mechanism can be attributed to the approximation of the macro-modelling approach which treats masonry as a continuum.

The mesh sensitivity of the numerical response is demonstrated in fig. 5 both for multiple-leaf and solid wall panels. Two mesh sizes were considered, refined by factors 0.5 and 2. The maximum difference in terms of collapse load is approximately 3% for multiple-leaf wall; the maximum difference in peak strength is about 11% for solid wall. Besides, the predicted failure mechanism remains the same. In summary, the predicted results can be considered mesh independent for practical purposes.

4 Conclusions

Numerical simulations of the experimental response of two different types of tuff masonry panels subjected to monotonic shear loading have been performed. A refined anisotropic continuum model is adopted and the interpretation of results has been carried out using DIANA FEM code, version 8.1. It is shown that, if a macro-modelling strategy can be applied, the proposed anisotropic model is able to predict well the behaviour of masonry structures, as well as sufficiently accurate post-peak response and failure mechanisms, even if some aspects need further optimizations. For the non-linear analysis of tuff masonry structures, calibration of relevant parameters is carried out and correlations between relevant mechanical parameters for both multiple-leaf and solid panels is provided. This is certainly of interest since experimental databases are generally poor and does not lead to fully satisfactory code provisions and retrofit design. Experimental and numerical results can be therefore considered an additional step in the perspective of full knowledge in the behaviour of tuff masonry structures.

References

[1] Gruppo Nazionale per la Difesa dai Terremoti GNDT, Censimento di vulnerabilità a campione dell'edilizia corrente dei Centri abitati, nelle regioni Abruzzo, Basilicata, Calabria, Campania, Molise, Puglia e Sicilia. Progetto LSU, 2000 (in Italian).

[2] FEMA Prestandard and Commentary for the Seismic Rehabilitation of Buildings, Rep. FEMA 356, Washington, DC, Federal Emergency Management Agency, 2000.

[3] European Committee for Standardization (CEN 2003), Design provisions for earthquake resistance of structures – Part 1: General rules, seismic actions and rules for buildings, EN 1998-1, Eurocode 8, Brussels, Belgium, 2003.

[4] Ordinanza n. 3431. Ulteriori modifiche ed integrazioni all'ordinanza del Presidente del Consiglio dei Ministri. n. 3274 del 20 marzo 2003. OPCM 3431, Gazzetta Ufficiale n.107, 10 maggio 2005 (in Italian).

[5] Marcari, G., Prota, A., Manfredi, G. & Pecce, M., In-plane shear performance of masonry panels strengthened with FRP. *Composites Part B: Engineering*, **38(7-8)**. pp. 887-901, 2007.

[6] Prota, A., Marcari, G., Fabbrocino, G., Manfredi, G. & Aldea, C. Experimental in-plane behavior of tuff masonry strengthened with cementitious matrix–grid composites. *ASCE Journal of Composites for Construction*, **10(3)**, pp. 223-233, 2006.

[7] Marcari, G., *Rinforzo sismico di murature di tufo con materiali fibrorinforzati*. Tesi di Dottorato di Ricerca in Ingegneria delle Costruzioni, Università degli Studi di Napoli Federico II, Napoli, 2004.
[8] Lourenço, P.B., de Borst, R. & Rots, J.G., A plane stress softening plasticity model for orthotropic materials. *International Journal for Numerical Methods in Engineering*, **40**, pp. 4033-4057, 1997.
[9] Lourenço, P.B., Rots, J.G. & Blaauwendraad, J., Continuum model for masonry: Parameter estimation and validation, *ASCE Journal of Structural Engineering*, **124(6)**, pp. 642-652, 1998.
[10] DIANA 2002, Displacement analysis finite element software, version 8.1. TNO-Building Division, Delft, The Netherlands, 2002.
[11] Ganz, H. R., & Thürlimann, B., Tests on masonry walls under normal and shear loading. Report No. 7502-4, Institute of Structural Engineering, ETH Zurich, Zurich, Switzerland (in German), 1984.
[12] Hoffmann, G. & Schubert, P. Compressive strength pf masonry parallel to the bed joints. *Proc. 10th Int. Brick and Block Masonry Conf*, eds N.G. Shrive and A. Huizer, University of Calgary, Calgary, Alberta, Canada, pp. 1453-1462, 1994.
[13] Faella, G., Manfredi, G. & Realfonzo, R., Stress-Strain relationship for Tuff Masonry: Experimental Results and Analytical Formulations. *Masonry International*, **7(2)**, pp. 55-61, 1993.
[14] Faella, G., Manfredi, G. & Realfonzo, R., Experimental evaluation of mechanical properties of old tuff masonry subjected to axial loading. *Proc. IV Congr. Naz. ASSI.R.C.CO.*, Prato, Italia, pp. 172-179, 1992.
[15] Schubert, P. & Weschke, K., *Verformung und Risssicherheit von Mauerwerk, Mauerwerk-Kalender*, ed. Ernst & Sohn, Berlin, Germany, pp. 145-159, 1986.
[16] Tassios, T.P., *Meccanica delle murature.* Liguori Ed, Napoli, Italia, (in Italian), 1998.
[17] Pina-Henriques, J.L., Masonry under Compression: failure analysis and Long-Term Effects. Doctoral Thesis, Escola de Engenharia, Universidade do Minho, October 2005.
[18] European Committee for Standardization (CEN 2005), Design of masonry structures – Part 1-1: General rules for buildings – Rules for reinforced and unreinforced masonry: EN 1996-1-1. Eurocode 6, Brussels, Belgium, 2005.
[19] American Society for Testing Materials (ASTM 1981), *Standard test method for diagonal tension (shear) in masonry assemblages.* E519-81, Philadelphia, 1981.

Experimental evaluation of stone masonry walls with lime based mortar under vertical loads

M. Abdel-Mooty, A. Al Attar & M. El Tahawy
Faculty of Engineering, Cairo University, Egypt

Abstract

Stone masonry walls of natural limestone blocks and lime based mortar similar to those used in ancient structures are tested under vertical loading and presented in this paper. The research program includes testing of wall components as well as wall assemblages under vertical loads to study the different factors affecting the wall strength, ductility and load-deformation behaviour up to failure. The considered factors cover: natural limestone block strength, lime based mortar strength, the relative dimensions and composition of the wall assembly as well as the thickness of mortar joints. Two types of natural stones of different strength, several mixes of lime based mortar with different strengths, and different systems and composition of wall models are tested in this research. Twelve wall specimens are considered including six single leaf walls, five double leaf walls and one cavity wall with shear key. It was found that the type and strength of mortar has significant effect on the overall wall behaviour. Also of paramount importance is the effect of the thickness of mortar joints on the wall ductility. The stress distribution within the wall component was recorded during wall testing to failure for further analysis.
Keywords: limestone masonry wall, lime based mortar, wall ductility, testing.

1 Introduction

Structural restoration of historical architecture involves in many cases strengthening and rebuilding of stone masonry walls at designated locations to restore the integrity of the monument. This restoration effort must be made with minimum intervention and must be compatible with the old material of the wall, basically limestone blocks and lime based mortar [1, 2].

Limestone masonry with lime based mortar is a relatively complex system whose behaviour is, in general, nonlinear [3–5]. The wall structural performance under load and strength depends on many factors such as the nature of stone, dimensions of the single block, arrangement in the structure and also the type of mortar that is used to connect the stones. Reliable numerical modelling of the wall behaviour requires accurate determination of the mechanical properties of its constituencies and detailed calibration of the developed models against comprehensive experimental test results [6–8]. This research tries to obtain a reliable evaluation from an experimental programme on masonry wall components and wall system.

The structural behaviour of the stone masonry wall depends very much on the mechanical characteristics of its components, namely the stone and the mortar. Furthermore, the structure of the wall such as single leaf, double-leaf of cavity wall with rubble infill significantly affects its load capacity and failure mode. The above factors in addition to the thickness of the mortar joints are investigated in this research.

The materials used in this study are natural lime stone blocks cut from local regions in Egypt. Three types of lime based mortars were used in this research. The first type consists of sand, lime and water. The second type consists of lime, sand, water and small amount of white cement to give strength to the mortar. The third type consists of lime, sand, water, small amount of white cement and small amount of stone rubbles. These three types of mortar are similar to that used in ancient constructions. The amount of white cement was added to give higher strength to the mortar. Several mortar mixes were tested in this study. The mortar mixes are classified as weak mortar (type one), strong mortar (type two) and strong mortar with stone rubbles (type three).

2 Experimental program and test set-up

The test programme aimed at evaluating the effect of different parameters on the performance of limestone masonry walls with lime based mortar under in-plane vertical loads up to failure (table 1). The tested parameters are:

1. Wall structure where three different wall systems were used (fig. 1):
   a. Single wall model of 120 mm thickness,
   b. Double wall model of 250 mm thickness,
   c. Cavity wall model made of external leaf stone blocks of 120 mm thickness and lime based rubble infill of 210 mm thickness. Thus, the overall thickness of the wall is 450 mm.

2. Lime based mortar strength where two types of mortars were used:
   a. Weak lime mortar,
   b. Strong lime mortar with white cement of 5% lime weight.

3. Three different thicknesses of horizontal and vertical lime based mortar joints: 15, 10, and 5 mm.

Table 1: Tested wall specimens' dimensions and parameters.

Wall Type	Wall No.	Wall Length mm	Wall Height mm	Wall Thickness mm	Mortar Thickness mm	Mortar Type
Single wall model	W1	780	1050	120	15	Weak
	W2	780	1050	120	15	Strong
	W3	770	1000	120	10	Weak
	W4	770	1000	120	10	Strong
	W5	760	950	120	5	Weak
	W6	760	950	120	5	Strong
Double wall model	W7	780	1050	250	15	Strong
	W8	770	1000	250	10	Weak
	W9	770	1000	250	10	Strong
	W10	760	950	250	5	Weak
	W11	760	950	250	5	Strong
Cavity wall	W12	780	1050	450	15	Strong

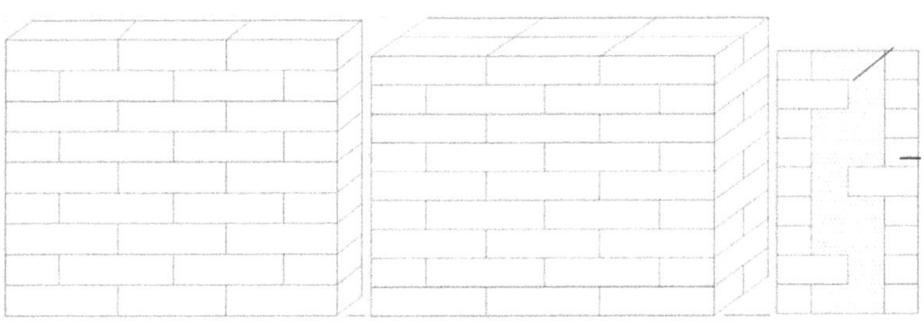

Single wall model Double wall model Cavity wall with infill

Figure 1: Tested models of wall specimens.

The effect of the above parameters on the performance of the limestone masonry walls was evaluated by examining the in-plane compressive failure load, load deformation relationship to failure, wall ductility, and cracking pattern and wall integrity at failure load.

Limestone blocks of dimensions 250×120×100 mm^3 cut from the Mokattam mountain quarry in Egypt were used in this research. The mechanical properties of the limestone block were experimentally determined. Four blocks were tested under compression yielding average compressive strength of 30.9 N/mm^2. The average of indirect splitting tension tests on three specimens was used to determine the tensile strength of the used limestone blocks as 1.3 N/mm^2.

In a previous comprehensive study by the first author, different types of lime based mortar for restoration of historical building were studied in detail where the advantages and limitations were identified [9]. In this paper, three types of lime based mortar were developed and tested. Those are designated as weak mortar of 25% lime and 75% sand; strong mortar of 25% lime, 5% white cement and 70% sand, and strong mortar with stone rubble of 25% lime, 5% white cement, 5% limestone rubble, and 65% sand. Forty four mortar cubes and six mortar cylinders were tested for the different mixes. The average 28 days compressive strength was experimentally identified as 0.14 N/mm^2 for weak mortar, 0.6 N/mm^2 for strong mortar and 0.77 N/mm^2 for strong mortar with rubbles.

All specimens were built on a reinforced concrete base beam to facilitate movement and placement of the specimens on the test frame after 28 days from construction. To ensure even distribution of compression loads on the top of wall models, a system of steel beams were used to distribute the point load of the actuator evenly on the wall (figs. 2 and 3). The capacity of the test frame is 5000 kN connected to a load cell of the same capacity. The instrumentation attached to the specimens also included LVDT to measure the vertical displacement of the wall; LVDT to measure the out of plane movement of the wall during testing, and strain gauges attached to the face of the stone blocks to measure vertical and horizontal strain (fig. 4). All specimens were loaded by

Figure 2: Experimental set-up.

Figure 3: Loading beam on wall top.

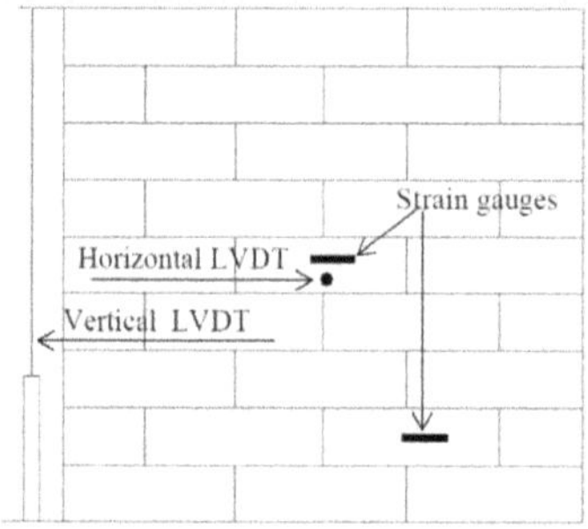

Figure 4: Instrumentation layout.

increasing the in-plane vertical load up to failure. The loads, vertical deformation, lateral deformation and strains were recorded during testing using a data acquisition system for further analysis.

3 Experimental results

The wall specimens were continuously loaded up to failure, and the data was recorded using a data acquisition system connected to a personal computer. In all tested specimens of single leaf wall models W1 to W6, it was observed that the initial cracking started from a stone above the vertical joint and extended up to the top of the wall as shown in fig. 5. The average stress on the wall section was calculated by dividing the applied load by the gross cross sectional area of the wall. Figs. 6–11 show the variation of average stress with the average strain up to failure for single leaf wall models. The values of failure loads and the corresponding stresses for all walls are summarized in table 2.

Wall W1 Wall W2 Wall W3

Figure 5: Crack pattern at failure of single leaf wall structure.

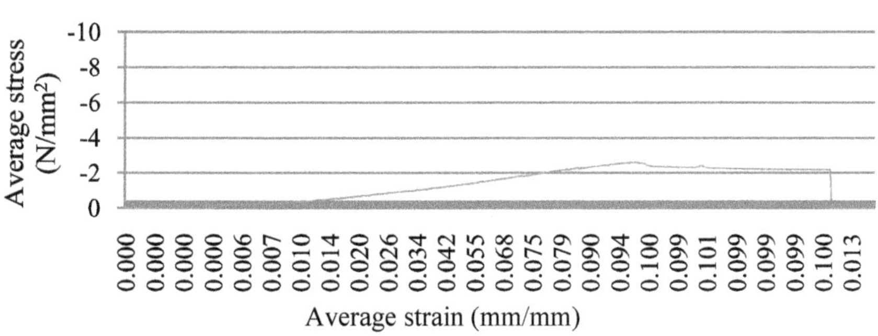

Figure 6: Stress-strain relationship of specimen W1 to failure.

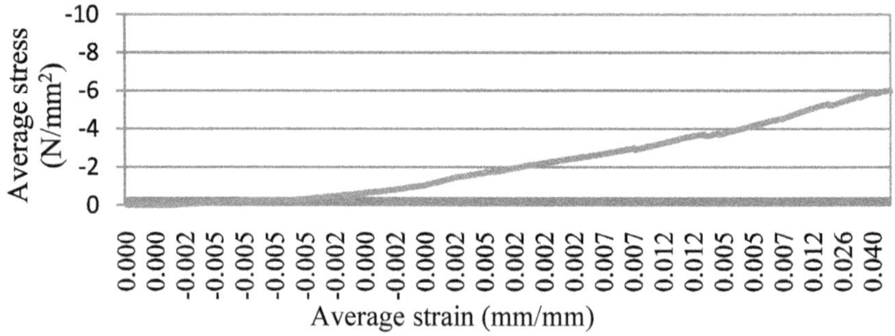

Figure 7: Stress-strain relationship of specimen W2 to failure.

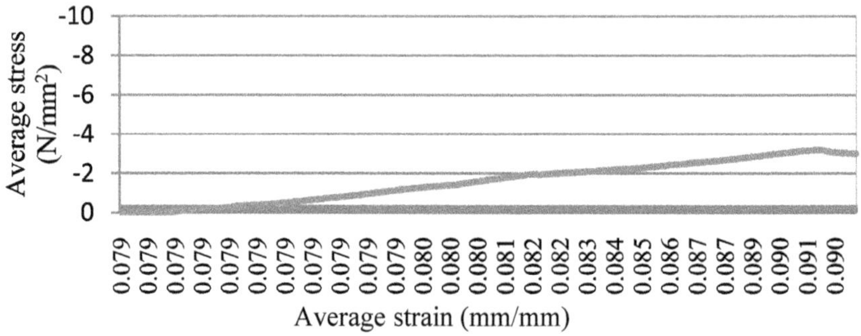

Figure 8: Stress-strain relationship of specimen W3 to failure.

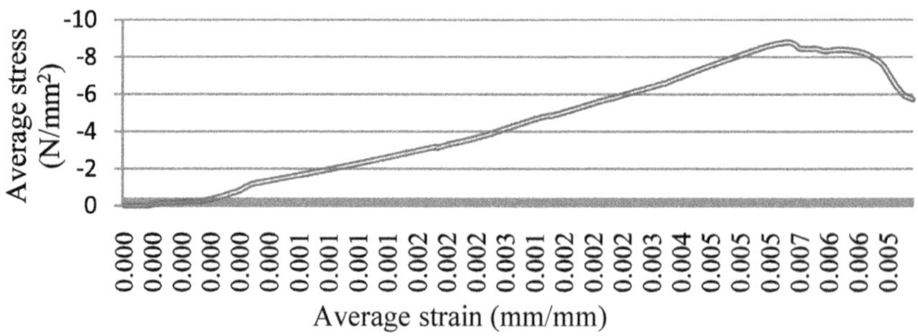

Figure 9: Stress-strain relationship of specimen W4 to failure.

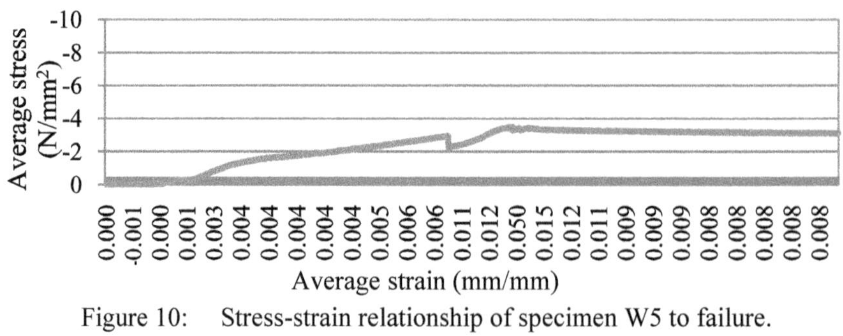

Figure 10: Stress-strain relationship of specimen W5 to failure.

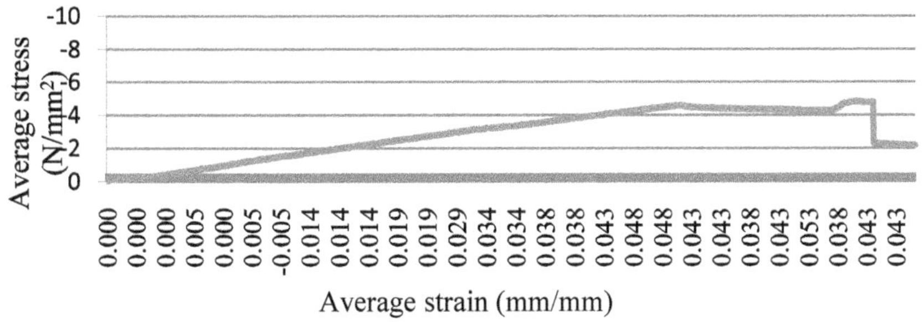

Figure 11: Stress-strain relationship of specimen W6 to failure.

Wall W7 Wall W9 Wall W11

Figure 12: Typical failure of double leaf stone masonry wall specimen.

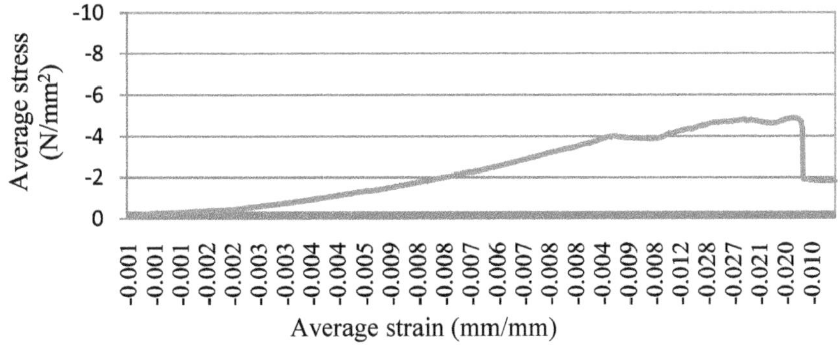

Figure 13: Stress-strain relationship of specimen W7 to failure.

Double leaf wall models were tested under increasing in-plane compressive load to failure. In almost all walls, cracking due to splitting of stone resulting from transversal stresses due to mortar squeezing started at the bed joint at the bottom of the wall and propagated upward. Eventually the wall failed due to splitting failure of the two faces of the wall as shown in fig. 12. Figs. 13 and 14 display the variation of the average compressive stress with the average strain up to failure for wall models W7 and W11, respectively. The results of the test for walls W7 to W11 are shown in table 2.

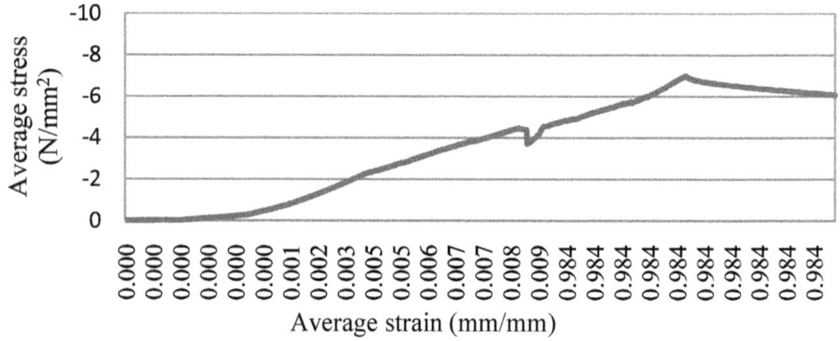

Figure 14: Stress-strain relationship of specimen W11 to failure.

Table 2: Summary of test results.

Wall Type	Wall No.	Mortar type	Mortar joint thickness mm	Failure load (kN)	Stress at failure N/mm^2
Single wall model	W1	Weak	15	240	2.6
	W2	Strong	15	590	6.3
	W3	Weak	10	360	3.9
	W4	Strong	10	900	9.7
	W5	Weak	5	350	3.8
	W6	Strong	5	456	5.0
Double wall model	W7	Strong	15	918	4.7
	W8	Weak	10	850	4.4
	W9	Strong	10	970	5.0
	W10	Weak	5	430	2.3
	W11	Strong	5	890	4.7
Cavity wall	W12	Strong	15	600	1.7

Wall W12 is a cavity wall (450 mm thickness) build with lime stone blocks 120 mm thick on the outer faces with 15 mm thick mortar joints. The wall faces are bonded together with 210 mm thick strong mortar with limestone rubble as infill between the wall faces. The specimen was continuously loaded up to failure load at 600 kN. The cracking started from the two side faces of the wall, and propagated upward. The mode of failure was mainly due to separation of the two faces of the cavity wall as shown in fig. 15. Cracks propagated around the stones instead of through stone blocks due to the weakness of the thick mortar infill compared to the stone blocks.

W12 during testing — Crack initiation in W12

Figure 15: Stress-strain relationship of specimen W7 to failure.

4 Analysis of experimental results

The effect of the different parameters on the wall performance under increasing load up to failure is presented in this section. Those parameters include the mortar strength, the mortar thickness and the wall system. For a fair comparison, since not all walls are of the same thickness and construction, the analysis is based on the average compressive strength instead of the failure load. Fig. 16 displays the failure stress of the tested twelve walls.

The existence of a small amount of white cement (5% of weight) in the lime mortar increases the mortar strength from 0.14 N/mm^2 to 0.6 N/mm^2. This results in significant increase in the strength of the walls as shown in fig. 17. This effect was evident for the different thicknesses of mortar joint.

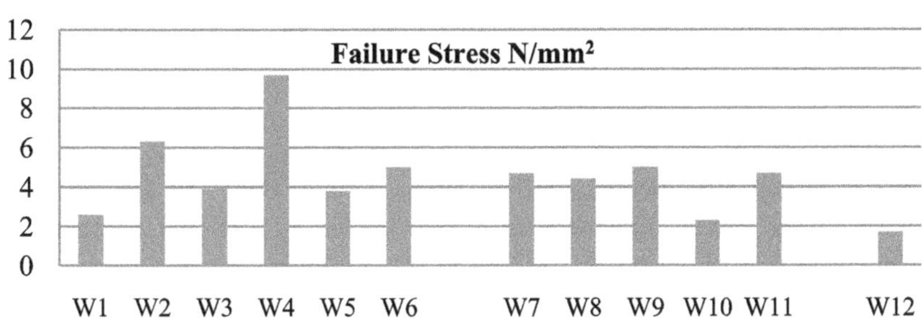

Figure 16: Comparison of the failure stress of the twelve tested walls.

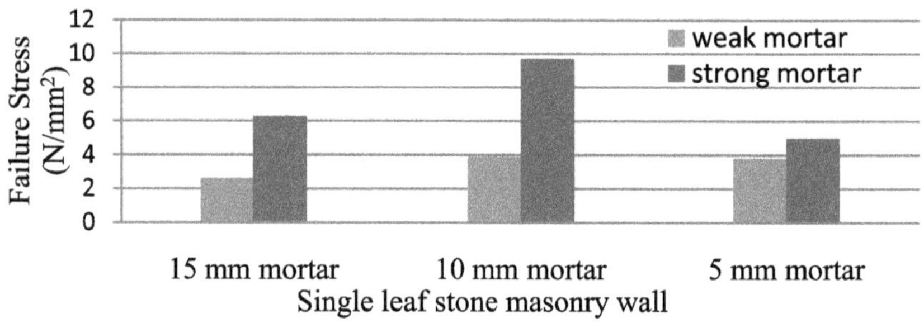

Figure 17: Effect of mortar strength on the strength of single leaf wall.

For single leaf walls W1 and W2 with thickness of mortar 15 mm, the strength of wall W2 with strong mortar is higher than that of W1 with weak mortar, almost 2.5 folds (150% increase). The same effect was observed for wall W3 and W4 of single walls with mortar thickness of 10 mm. This effect was less pronounced in walls W5 and W6 with mortar thickness of 5 mm where the increase was only 30%. Similar behaviour was observed in the double leaf walls.

Furthermore, the increase in the mortar strength results in significant increase in the overall stiffness of the wall. This effect is clearly shown in fig. 18 for wall W1 with weak mortar and wall W2 with strong mortar.

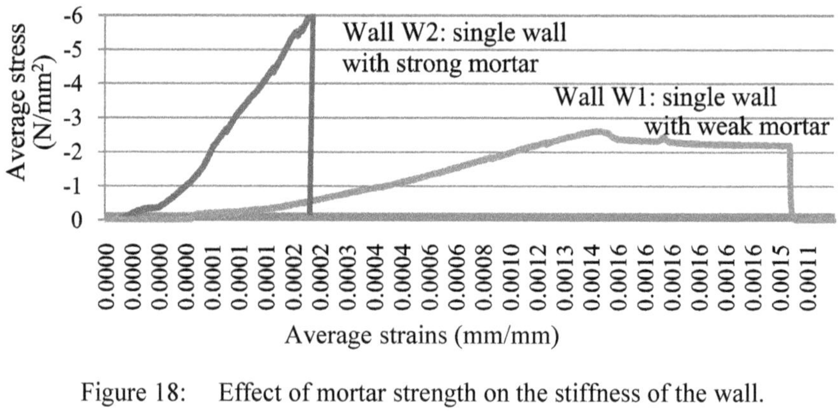

Figure 18: Effect of mortar strength on the stiffness of the wall.

The effect of the thickness of mortar joints on the overall strength of masonry walls systems is displayed in fig. 19. It is observed that the highest strength is attained at a mortar thickness of 10 mm. Increasing the mortar thickness to 15 mm slightly decreases the wall overall strength however improves the wall ductility as shown by the stress-strain curves of figs. 6–11, 13 and 14. Also, decreasing the mortar thickness to 5 mm decreases both the strength and ductility.

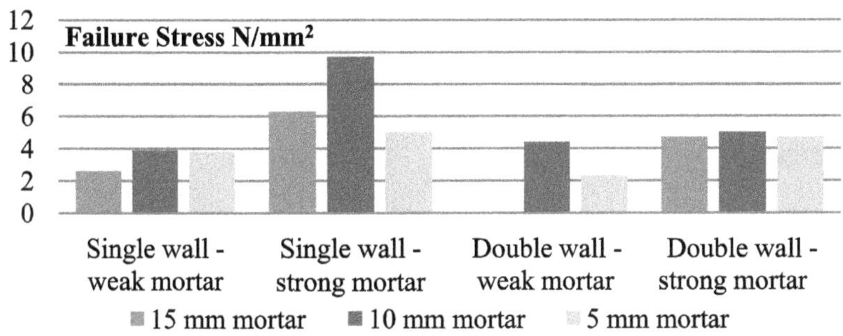

Figure 19: Effect of mortar joint thickness on the wall strength.

5 Summary and conclusions

Twelve limestone masonry wall models with lime based mortar were tested under increasing in-plane compressive load to failure. The effect of different parameters on the wall performance was evaluated in this study. Those parameters include: wall structure as single leaf, double leaf, and cavity wall with rubble infill; strength of lime based mortar where two types of mortar designated as weak and strong mortar are used; and the thickness of mortar joints where three thicknesses, 15 mm, 10 mm, and 5 mm, were tested. The mechanical properties of the natural stone units and mortar cubes were determined experimentally in this work.

The main modes of failure in almost all tested walls was splitting tension failure of stone blocks due to squeezing of mortar, followed by crushing of stones due to increasing compressive loads.

It is evident that, the strength of mortar in joints significantly affects the overall strength of the wall. Increasing mortar strength by adding white cement (5%) increases the overall strength of the masonry wall by almost 2.5 times.

Mortar is able to withstand higher compressive stresses in bed joints than stresses in cubes because of the confining restrains from the stone units. However, the lateral deformation of mortar is found to be greater than that of the stone blocks due to squeezing of mortar which causes lateral tension stresses in stone blocks and causes splitting of stones. Thus, the tensile strength of the stone blocks affects the overall compressive strength of masonry walls.

The behaviour of mortar joints in stone masonry walls under loading gives ductility to the wall and provides warning signs before collapse in the form of mortar squeeze and splitting cracks in stone blocks. It was found that walls built with thin mortar joints failed without sufficient warning compared with walls with thicker joints.

The thickness of mortar greatly affects the overall strength of the stone masonry wall. Increasing mortar thickness from 10 mm to 15 mm reduces the compressive strength by 29% for single walls, and 22% for double walls. Reducing mortar thickness from 10 mm to 5 mm reduces the compressive strength by 0.5% for single wall and 10% for double wall. In general, increasing mortar thickness reduces the wall strength but increases its ductility and the highest strength of the wall was achieved at mortar joint thickness of 10 mm.

References

[1] Abdel-Mooty, M.N., Mourad, S.A. & Abdel-Gawad, A.A., Structural adequacy, damage assessment, and repair methodology of historical walls. *Arab Conference for Strengthening and Rehabilitation of Structures*, Cairo, Egypt, Vol. 2, pp. 1041-1057, 1998.

[2] Hayen, R., Van Balen, K., Van Gemert, D. & Schueremans, L., The influence of the brick/mortar interaction on the mechanical behaviour of historic masonry. *STREMAH VI Computer Methods in Structural Masonry,* pp. 22-24, 2003.

[3] Abdel-Mooty, M.N. & Abdo, S.M., Structural behaviour of natural stone masonry walls under vertical and lateral loads. *International Conference on Future Vision and Challenges for Urban Development*, Cairo, Egypt, pp. 20-22, 2004.

[4] Hayen, R., Schueremans, L., Van Balen, K. and Van Gemert, D., Triaxial testing of historic masonry, test set-up and first results. *Structural Studies, repairs, and Maintenance of Historical Buildings, VII*, ed. C.A. Brebbia, WIT Press: Southampton, pp. 151-160, 2001.

[5] Valluzzi, M. R., Porto, D. A. & Modena, C., Behavior of multi-leaf stone masonry walls strengthened by different intervention techniques. *Historical Constructions*, eds. P.B. Lourenco & P. Roca, Guimaraes, pp. 1023-1025, 2001.

[6] Bartoli, G., Casamaggi, C. & Spinelli, P., Numerical modelling and analysis of monumental buildings: a case study. *Memoria Presentata al Convegno CICOP "FIRENZE2000", 5th International Congress on Restoration of Architectural Heritage*, 2000.

[7] Pina-Henriques, J.L., Lourenço, P.B., Binda, L. & Anzani, A., Testing and modelling of multiple-leaf masonry walls under shear and compression. *Proc. 4th Int. Seminar on Structural Analysis of Historical Constructions*, Padova, Italy, pp. 299-310, 2004.

[8] Fahmy, E. & Ghoneim, T., Behavior of concrete block masonry prisms under axial compression. *Canadian Journal of Civil Engineering,* **22**, pp. 898-915, 1996.

[9] Abdel-Mooty, M., Khedr, S. & Mahfouz, T., Evaluation of lime mortars for repair of historic structures. *11th Int. Conf. on Structural Repairs and Maintenance of Heritage Architecture*, Tallinn, Estonia, 2009.

Effective hygric and thermal parameters of historical masonry accessed on effective media theory principles

Z. Pavlík, E. Vejmelková, L. Fiala, M. Pavlíková & R. Černý
Czech Technical University in Prague, Czech Republic

Abstract

The experimental determination of hygric and thermal properties of two types of historical masonry stone materials and lime-metakaolin mortar is initially carried out. Nominally, water vapour transport properties, sorption isotherms, thermal conductivity with its dependence on rising moisture, and basic physical properties are measured using sophisticated laboratory methods. The researched materials are usually applied in the Czech territory for restoration and reconstruction of historical masonry and other structures. The measured material properties are subjected to a homogenization procedure, and the effective heat and moisture transport and storage parameters of a typical fragment of masonry are calculated. For the homogenization, the mixing model originally derived for application in dielectric studies is employed taking into account the theoretical bounds of investigated material parameters. The obtained data give information on hygrothermal performance of the modelled masonry fragment and can find use, for example, in the damage assessment of historical buildings using methods of computational analysis.
Keywords: hygric properties, thermal properties, historical stone masonry, argillite, sandstone, lime-based plaster, homogenization techniques.

1 Introduction

The hygrothermal behaviour of climatically exposed components and structures of historical buildings is related to the hygric, thermal, mechanical and other physical and chemical properties of inbuilt materials. Their knowledge constitutes necessary information for the understanding of the building performance as well as a first step towards avoiding damage, degradation or the undue heat loss from the structures.

The damage assessment of historical masonry and other structures due to the negative effects of moisture and temperature can be carried out effectively by means of mathematical and computational modelling. In this way, the development of moisture and temperature fields with time can be obtained; this is crucial for a proper assessment of possible future damage [1]. Their prediction is a very important task when preserving historical bridges and buildings or insulating existing buildings and components. The moisture and temperature values can be then assigned to the mechanical properties and to the risk of consequent damage. For instance, Charles Bridge in Prague, Czech Republic, which was subjected to reconstruction work lately, is a typical example of extensive damage mostly brought about by temperature and moisture influences.

For the effective application of computational modelling to a buildings' performance analysis, a complete knowledge of the properties of the applied materials as well as the initial and boundary conditions of the simulations is necessary. As the initial conditions for the hygric and thermal analysis, the moisture and temperature fields experimentally measured from the investigated structure can be used. The boundary conditions of the computer simulations are given by the climatic loading that can be simply adopted using meteorological data.

In engineering practice, simplified analyses of building structures' behaviour are very often used. Here, in the computational simulations of heat and moisture transport processes, brick or stone masonry is often understood as a single material, the brick or stone itself. This simplification may work reasonably well when the properties of the brick or stone and the mortar are similar. However, this may not always be the case in some historical buildings, where low-quality mortars were often used [2]. On that account, the results of such calculations deviate from the real state of the buildings, and cannot be effectively used for assessing some appropriate materials' damage and problems in hygrothermal behaviour. Therefore, a solution of this problem must be found in order to improve the accuracy and reliability of computational modelling of building structures performance.

There are two ways of meeting the above given requirements for reliability and accuracy in computational simulations of heat and moisture transport. The first possibility is the use of sophisticated simulation tools that allow discretization of the studied structure with such a resolution, that all the properties of inbuilt materials can be assigned values corresponding to specific places of the structure. Although this procedure is generally correct, it is too complex and sophisticated for wider application in engineering practice, especially in the case of conservationists of historical buildings who are usually well informed on materials' problems without knowledge of computational modelling methods.

The second, and probably the most straightforward way to solve the problem of variations in the properties of the particular components of masonry is the utilization of the effective media theory. This procedure is introduced in the presented work as an effective tool for the simplification of computational modelling of transport processes in the masonry. The application of homogenization principles produces the macroscopic equations which may be used when analyzing the masonry as a whole. In these equations, the effective parameters are used instead of the parameters of the brick or stone.

It should be quite apparent that the homogenization process cannot be carried out without the exact knowledge of the properties of all materials constituting the masonry and of the amount of each material in the analyzed wall. Therefore, the experimental measurement of all the material parameters characterising the process of heat and moisture transport and accumulation in inbuilt masonry materials is first necessary. These data are then introduced in the mixing models based on effective media theory, and the effective thermal and hygric parameters of the whole masonry can be calculated.

2 Studied materials

Two types of stones that were often used in Czech region in the past as bearing masonry materials are analysed together with lime-based mortar with pozzolana admixture.

Many historical buildings in the Czech Republic were built using similar kinds of sandstone. Siliceous raw-grained sandstone was usually used for historical architectural constructions (walls, portals, window frames) for its strength. Ornamental parts of the architecture (gothic flowers, romantic shells) and sculptures (from the Romanesque period up to now) were made of fine-grained calcite-argillaceous sandstone. In this work, sandstone from Mšené-lázně quarry, Czech Republic, is chosen. It is fine-grained psamitic equigranular rock, about 95% of which is made up of suboval quartz clasts. Other mineral grains are present only as minorities (tourmaline, epidote, muscovite and zircone). Quartz grains reach up to 0.1 mm in diameter, but those of muscovite are larger, up to 0.3 mm [3]. The matrix is formed by clay minerals (mainly kaolinite).

Also the argillite was very popular material in historical architecture. It was used for sacral as well as for secular buildings, flagstone pavements, roof slabs, and facing. The studied argillite is coming from quarry Džbán, Czech Republic. Its main constituents are illite, calcite, minerals on the basis of SiO_2 having granularity 0.3–0.15 mm, feldspar, and mica, whereas rigid materials form 40–60% of argillite volume.

Within the reconstruction of historical buildings, renewal and restoration of interior and exterior plasters and mortars is often carried out. From the point of view of a historian, it is not acceptable to use lime-cement mortars in Romanesque, Gothic, Renaissance, and Baroque buildings. However, the pure lime-based plaster does not exhibit sufficient resistivity against moisture action. On that account, proper hydraulic admixtures must be used that enhance the durability of mortars. These materials must have similar composition as the historical materials and they have to be applicable by the original technological processes.

As the chemical analyses of many plasters from historical buildings show, the external plasters from past centuries that are preserved until today contain products formed by lime reaction with pozzolanic or hydraulic admixtures. Pozzolanic admixtures appeared to have a positive effect on properties of lime binder in the past. On that account, a lime-based mortar with metakaolin addition as pozzolana admixture was chosen. The composition of the lime-metakaolin plaster was as follows: hydrated lime – 400 g, metakaolin – 80 g, natural quartz sand with continuous granulometry 0 to 2 mm – 1 440 g, and water – 480 g.

3 Experimental methods

Basic material properties of all tested materials were determined at first. Bulk density and matrix density were measured using the gravimetric method and helium pycnometry, and then the total open porosity was calculated. The samples' dimensions for these measurements were 40×40×20 mm^3.

For the determination of water vapour transmission properties, the cup method was applied, according to the European standard EN ISO 12571 [5]. The measurements were carried out under steady state isothermal conditions. The dry cup arrangement was used in the experiment. Here, the sealed cup with the studied material sample containing burnt $CaCl_2$ (0% relative humidity) was placed in a controlled climatic chamber at 25±0.5°C and 50% relative humidity and it was weighed periodically. The circular samples had a diameter of 95 mm and a thickness of 20 mm. The steady state values of mass gain were utilized for the determination of the water vapour permeability. On the basis of measured water vapour permeability, the water vapour diffusion coefficient and water vapour resistance factor were calculated using the simple formulas given in [6, 7].

For the determination of moisture diffusivity as function of moisture, the moisture profiles were measured at first. The measurements were carried out on samples having dimensions of 40×20×300 mm^3 in 1-D experimental arrangement of water transport. The moisture content at a specific position in the sample was measured by the capacitance technique calibrated by the gravimetric method. The moisture dependent moisture diffusivity was then calculated using the inverse analysis of the measured moisture profiles by means of the Boltzmann-Matano treatment [8].

In the sorption isotherm measurement, the samples were placed into the desiccators with different salt solutions to simulate different values of relative humidity. The mass of samples was measured in specified periods of time until a steady state value of mass was achieved. Then, the volumetric moisture content was calculated and the sorption isotherm of each tested material was plotted.

The thermal conductivity as the main parameter of heat transport was determined using the commercial device ISOMET 2104 (Applied Precision, Ltd.). ISOMET 2104 is a multifunctional instrument for measuring thermo-physical parameters which is based on the application of an impulse technique and is equipped with various types of optional probes [9]. Thermal conductivity was measured in the moisture range from the dry state to full water saturation on the 70 mm cubes.

4 Effective media theory and homogenization techniques

The application of the effective media theory allows the determination of the properties of the whole masonry or building structure instead of the parameters of the particular inbuilt materials. In terms of effective media theory, the final composite structure (in our case stone masonry wall) can be considered basically as a mixture of walling blocks and mortar. In more precise calculations, each of these materials can be further considered as a mixture of three phases, namely solid, liquid and gaseous phase (in four phase systems, the effect of bound water can be included) that form their matrix and porous space. There are two basic approaches that can be applied for determination of thermal and hygric properties of the stone masonry. The first possibility is to apply homogenization techniques on the moisture dependent material data of the materials encountered in the masonry. This simplified procedure was used in this work. The second possibility is based on the complete knowledge of material properties of the particular components forming the porous body of the structure (dry stone, dry mortar, free water, bound water and air). From the properties of particular components and their volumetric fractions the effective properties of the stone masonry can be assessed.

On the basis of previous experience with application of homogenization techniques, the original Lichtenecker's formula [10] was chosen for the calculations performed in the presented work; this formula was proved to be satisfactory for the evaluation of moisture dependent thermal and hygric properties of porous building materials [11]. Also the application of four phase models looks promising. However, in case of the masonry studied in this work, the determination of the amount of bound water represents a very complex problem.

The Lichtenecker's equation adjusted for the studied masonry wall assumes that the effective hygric and thermal parameter of the considered material satisfies the equation

$$p_{eff}^{k} = f_b p_b^k + f_m p_m^k , \tag{1}$$

where p_{eff} is the calculated effective parameter of masonry, f_b the volumetric fraction of walling blocks in masonry, f_m the volumetric fraction of mortar, p_b the measured material parameter of walling block, p_m the measured parameter of mortar, and k is a free parameter describing basically the path from the anisotropy at $k = -1.0$ to another anisotropy at $k = 1.0$. However, the Lichtenecker's equation may also be applied for isotropic materials. On the basis of previous experience, the value of the parameter $k = 0$ was used.

The effective parameters of a multi-phase material cannot exceed the bounds given by the parameters of particular fractions of its constituents. Here, the Wiener bounds according to the Wiener's original work were used [10, 11]. These bounds can be expressed by the following relations.

$$p_{eff} = \frac{1}{\sum_{i=1}^{n} \frac{f_i}{p_i}}, \tag{2}$$

$$p_{eff} = \sum_{i=1}^{n} f_i p_i , \tag{3}$$

where eqn (2) represents the lower limit and eqn (3) the upper limit of the investigated effective material parameter (f_i is the volumetric fraction of the particular phase, in our case argillite, sandstone and mortar), p_i its material parameter).

5 Studied masonry

For the application of the homogenization theory for the evaluation of the effective hygric and thermal parameters of masonry, a typical fragment of the wall was constructed in a simplified manner, similar to common brick masonry as shown in Figure 1.

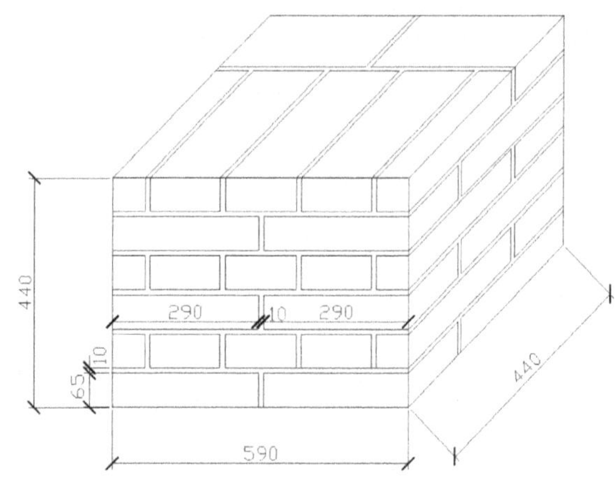

Figure 1: Scheme of the reference masonry wall (dimensions in mm).

Using this scheme, the volumetric fractions of the particular materials in the wall were calculated. The volumetric fraction of walling materials is equal to 0.824, whereas the volumetric participation of mortar is 0.176. Through the calculations, the effective parameters of two different walls were assessed. The studied walls always consisted of lime based metakaolin mortar, and of one walling material, nominally sandstone or argillite.

6 Results and discussion

The basic properties of all tested materials are summarized in table 1. Each result represents the average of five measured values. All the studied materials have shown high porosity; this is a very positive factor from the point of view of their presumed application in reconstruction of historical buildings.

The water vapour transmission properties of masonry materials are presented in table 2. Very low values of the water vapour resistance factor can be systematically seen; this is again a very promising finding for the application of all tested materials in historical masonry. Since a historical masonry usually exhibits increased moisture content, it is necessary to apply within the reconstruction processes such materials that will allow moisture evaporation from the renovated structures within the warm periods of the year.

Table 1: Basic material properties of masonry materials.

material	bulk density (kg/m^3)	matrix density (kg/m^3)	total open porosity (-)
sandstone	1807	2627	0.31
argillite	1353	2235	0.39
lime-metakaolin mortar	1690	2620	0.35

Table 2: Water vapour transmission properties of masonry materials.

	dry cup arrangement, 0–50%		
material	water vapour permeability (s)	water vapour diffusion coefficient (m^2/s)	water vapour resistance factor (-)
sandstone	2.4E-11	3.3E-06	7.0
argillite	2.9E-11	4.1E-06	5.7
lime-metakaolin mortar	1.8E-11	2.5E-6	10.1

Table 3a: Water vapour transmission properties of stone masonry.

Sandstone wall – homogenization, dry cup arrangement – 0–50% RH			
	water vapour permeability (s)	water vapour diffusion coefficient (m^2/s)	water vapour resistance factor (-)
Wiener's lower bound	2.29E-11	3.11E-06	7.40
Wiener's upper bound	2.24E-11	3.05E-06	7.55
Lichtenecker model, k=0	2.27E-11	3.08E-06	7.47

The effective diffusion parameters of the studied stone masonry, calculated on homogenization principles, are presented in tables 3a and 3b. For verification of the calculated results, also the limiting bounds of effective parameters are presented. Looking at the obtained data of water vapour transmission properties, it can be seen that all the effective parameters lie between the Wiener's bounds; this basically confirms their validity.

Table 3b: Water vapour transmission properties of stone masonry.

Argillite wall – homogenization, wet cup arrangement – 0–50% RH			
	water vapour permeability (s)	water vapour diffusion coefficient (m^2/s)	water vapour resistance factor (-)
Wiener's lower bound	2.74E-11	3.73E-06	6.17
Wiener's upper bound	2.61E-11	3.55E-06	6.47
Lichtenecker model, k=0	2.69E-11	3.65E-06	6.30

The effective moisture diffusivity as function of moisture content is given in figs. 2 and 3, which show the high dependence of this moisture transport parameter on moisture content. This materials' behaviour significantly affects the hygrothermal performance and consequently the durability. Sandstone systematically exhibits the highest moisture diffusivity. Its highly porous structure formed by high radius pores allows fast liquid moisture transport in comparison with argillite that is also characterised by high total open porosity, but its pore size is much smaller and structure more fine-grained.

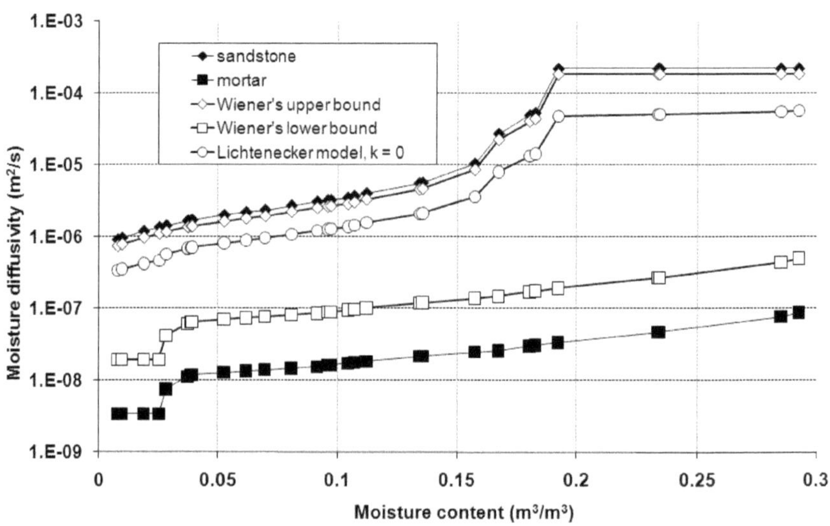

Figure 2: Effective moisture diffusivity of sandstone masonry.

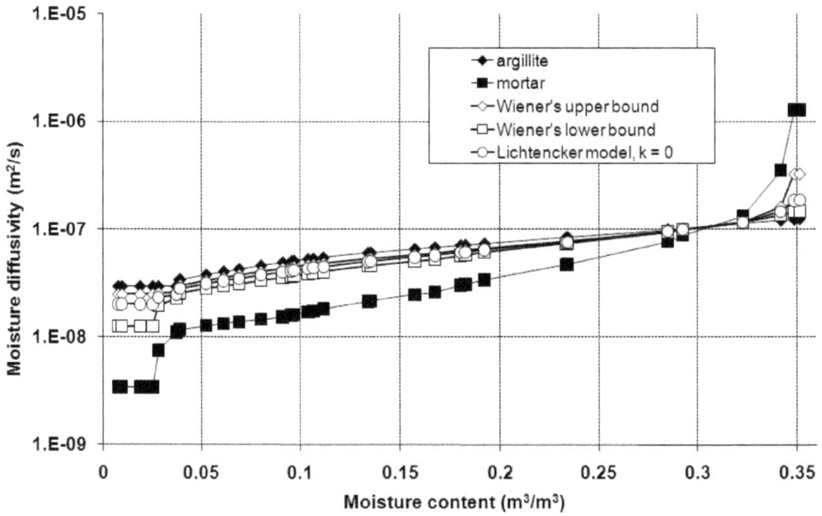

Figure 3: Effective moisture diffusivity of argillite masonry.

The calculated effective sorption isotherms of the studied masonry are given in figs. 4 and 5. Typically, the highest capacity for water vapour adsorption can be observed in lime-metakaolin mortar. However, in the hygroscopic moisture range, the highest values of adsorbed moisture were measured in argillite. The data obtained for sandstone were systematically the lowest.

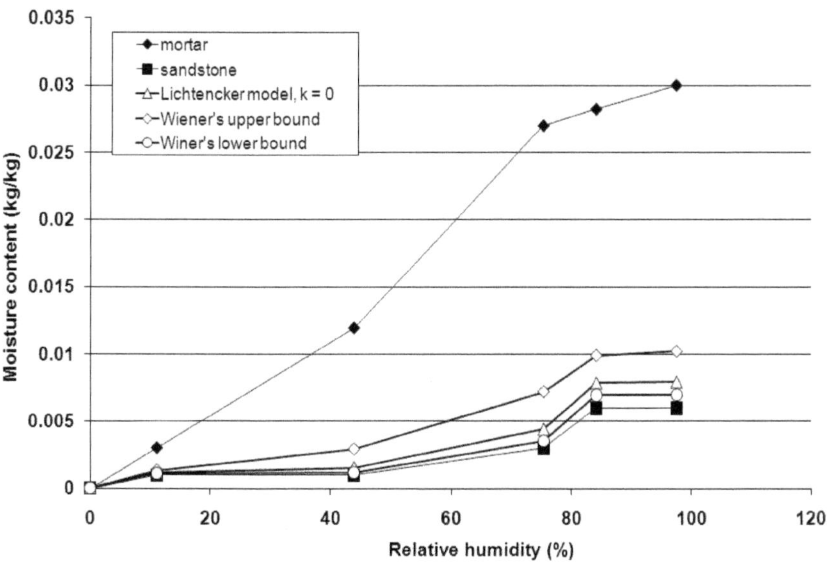

Figure 4: Effective sorption isotherm of sandstone masonry.

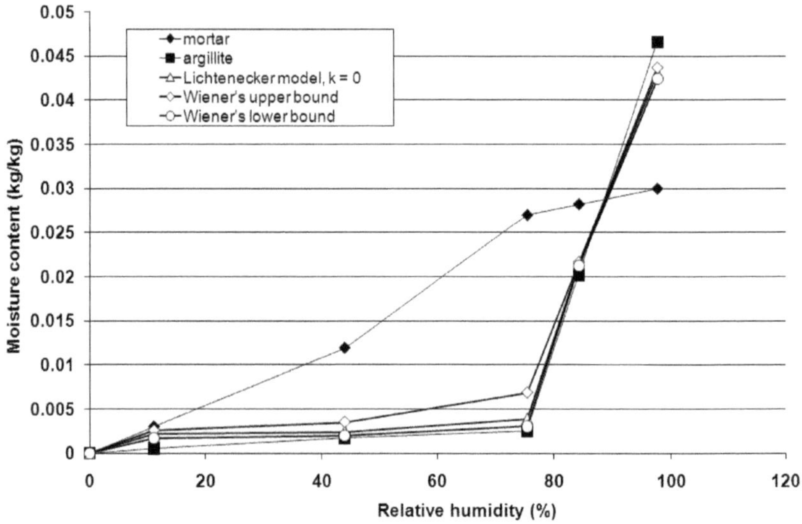

Figure 5: Effective sorption isotherm of argillite masonry.

The measured and calculated thermal conductivity of all studied materials and the two types of masonry is presented in figs. 6 and 7. The variation of the effective thermal conductivity with moisture content is systematically within the range of Wiener's bounds; this basically proves the reliability of the applied homogenization technique.

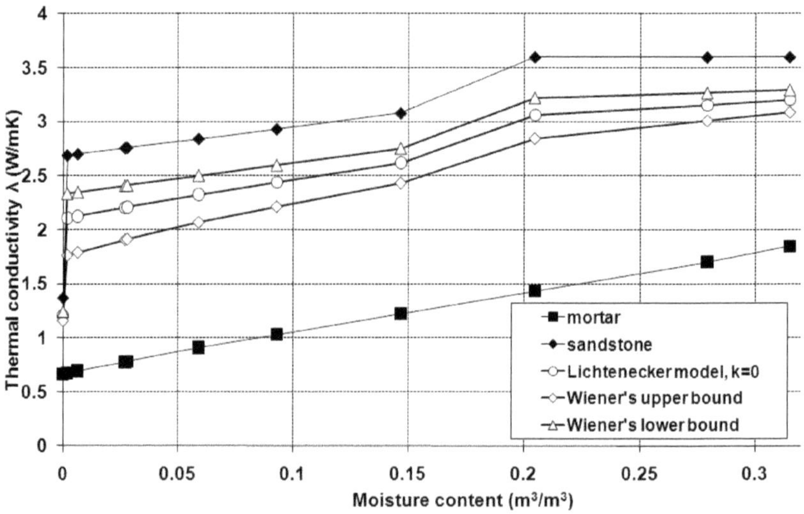

Figure 6: Effective thermal conductivity of sandstone wall.

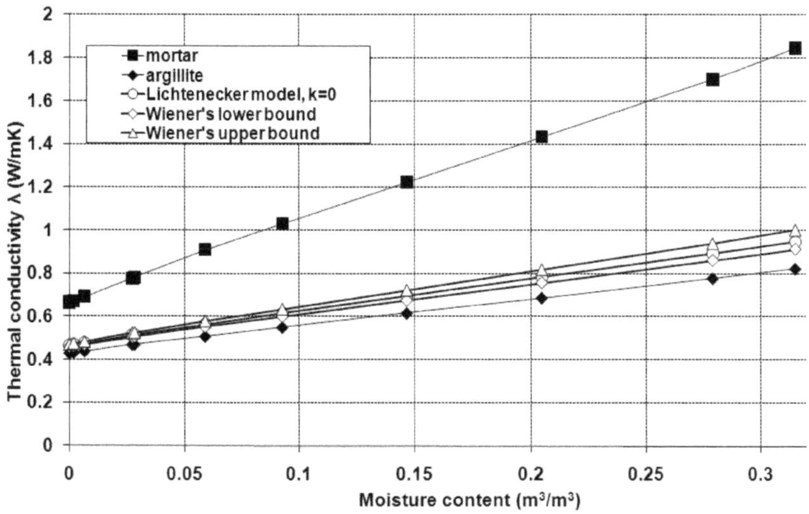

Figure 7: Effective thermal conductivity of argillite wall.

7 Conclusions

An example of application of the homogenization technique for the determination of hygric and thermal parameters of stone masonry was introduced in this paper. The obtained results indicate that the Lichtenecker's equation may be successfully used for such a type of applications. Nevertheless, there is still an unexplored task to verify the reliability of the applied homogenization model by laboratory experiments that should confirm or contradict the presented results.

Acknowledgment

This research has been supported by the Czech Science Foundation, under project No P105/12/G059.

References

[1] Pavlíková, M., Pavlík, Z. & Černý, R., Hygric and thermal properties of materials used in historical masonry. *Proc. of the 8th Symposium on Building Physics in the Nordic Countries*, Technical University of Denmark: Lyngby, pp. 903–910, 2008.

[2] Vejmelková, E., Pavlík, Z., Fiala, L., Pavlíková, M. & Černý, R., Heat and moisture transport properties of stone masonry materials. *Proc. of the Building Physics Symposium*, Katolieke Universiteit Leuven, Leuven, pp. 113–116, 2008.

[3] Pavlík, Z., Michálek, P., Pavlíková, M., Kopecká, I., Maxová, I. & Černý, R., Water and salt transport and storage properties of Mšené sandstone. *Construction and Building Materials*, **22(22)**, pp. 1736–1748, 2008.
[4] Pavlík, Z., *Identification of Parameters describing the Coupled Moisture and Salt Transport in Porous Building Materials*, CTU Prague, p. 145, 2009.
[5] EN ISO 12572, *Hygrothermal Performance of Building Materials and Products, Determination of Water Vapour Transmission Properties*, European Committee for Standardization, Brussels, 2001.
[6] Jiřičková M., *Application of TDR Microprobes, Minitensiometry and Minihygrometry to the Determination of Moisture Transport and Moisture Storage Parameters of Building Materials*, CTU Prague, p. 102, 2004.
[7] Roels, S., Carmeliet, J., Hens, H., Adan, O., Brocken, H., Černý R., Pavlík Z., Hall, C., Kumaran, K., Pel, L. & Plagge, R., Interlaboratory comparison of hygric properties of porous building materials. *Journal of Thermal Envelope & Building Science*, **27(4)**, pp. 307–325, 2004.
[8] Roels, S., Carmeliet, J., Hens, H., Adan, O., Brocken, H., Černý, R., Pavlík, Z., Hall, C., Kumaran, K. & Pel, L., A comparison of different techniques to quantify moisture content profiles in porous building materials. *Journal of Thermal Envelope & Building Science*, **27(4)**, pp. 261–276, 2004.
[9] Jiřičková, M., Pavlík, Z., Fiala, L. & Černý, R., Thermal conductivity of mineral wool materials partially saturated by water. *International Journal of Thermophysics*, **27(4)**, pp. 1214–1227, 2006.
[10] Mňahončáková, E., Jiřičková, M., Pavlík, Z., Fiala, L. & Černý, R., Effect of moisture on the thermal conductivity of a cementitious composite. International Journal of Thermophysics, 27(4), pp. 1228–1240, 2005.
[11] Pavlík, Z., Vejmelková, E., Fiala, L. & Černý, R., Effect of moisture on thermal conductivity of lime-based composites. *International Journal of Thermophysics*, **30(6)**, pp. 1999–2014, 2009.

Investigation of commercial ready-mixed mortars for architectural heritage

D. Gulotta[1], L. Toniolo[1], L. Binda[2], C. Tedeschi[2],
R. P. J. van Hees[3,4] & T. G. Nijland[3]
[1]*Dipartimento di Chimica, Materiali e Ingegneria Chimica "Giulio Natta", Politecnico di Milano, Italy*
[2]*Dipartimento di Ingegneria Strutturale, Politecnico di Milano, Italy*
[3]*TNO Built Environment and Geosciences, The Netherlands*
[4]*Faculty of Architecture, Delft University of Technology, The Netherlands*

Abstract

In conservation practice, operations aimed at restoring the structural functionality of historical buildings are of primary relevance. Partial rebuilding, substitution of damaged elements and integration of deteriorated joints are interventions which require the introduction of new materials in heterogeneous and aged masonry systems. An appropriate durability of these interventions can only be assured if a deep knowledge of the original materials, as well as of their state of conservation, is combined with a reliable understanding of the behaviour of the new ones. The present paper reports the preliminary results of the investigation of four commercial ready-mixed mortars based on NHL binders. The initial composition of the products was assessed by means of XRD and FTIR. The hardened mortars characteristics were evaluated and their mechanical behaviour was tested as well. Though supposedly comparable products, the mortars showed remarkable differences for what concerning compositional features, microstructures, presence of additives and mechanical strength. All these aspects are crucial in order to evaluate their compatibility with respect to the historic materials and should be properly considered for the selection of the most appropriate product for each single case. Moreover, as they are often neglected or, at least, only partially stated in the technical data sheet provided by the manufacturers, they need to be determined prior to the application.
Keywords: ready-mixed mortars, mortar characterization, NHL, XRD, FTIR.

1 Introduction

A number of commercial ready-mixed mortars designed for restoration purposes are currently available on the market and their use in the conservation field is increasing. The diffusion of such products is mainly related to their ease of use, which also determines a generally higher cost respect to a manually prepared traditional mortar mix. Ready-mixed mortars are available as packed powder material composed of binder, aggregates of different nature and granulometry and additives already mixed together in appropriate ratios. The technical data sheets state the precise amount of water to be added and often supply indications about the mixing procedure. In such way, mortar's preparation is clearly a quicker operation compared to the traditional method and it can be easily performed also by unskilled labour.

Natural hydraulic lime is one of the most common binders used in commercial mixes for restoration. Technical standard UNI EN 459-1 [1] gives a definition for NHL (natural hydraulic lime) based on raw materials, production process, calcium hydroxide content and final mechanical behaviour. Natural hydraulic lime results from the burning of argillaceous or siliceous limestone, followed by reduction to powder by slaking, with or without grinding. NHL are able to set and harden even under water through hydration of the hydraulic compounds. Atmospheric carbon dioxide contributes to the hardening process, allowing the carbonation of calcium hydroxide. The burning temperature is kept below the clinkering point so that di-calcium silicate (C_2S) is the dominant hydraulic phase together with some gehlenite (C_2AS) and only low amount of highly reactive tri-calcium silicate (C_3S) typical of Portland cement can form [2].

Despite the increasing diffusion of NHL based ready-mixed mortars, a thorough characterisation of commercial products is rarely available. Several studies have been conducted in the past in order to develop methodologies for the characterization of ancient mortars and to evaluate the overall compatibility of the new materials introduced during the restoration operations [3–5]. Recent studies have also been focused on the characterisation of some XIX and XX century mortars [6, 7]. Methodologies have been defined in order to identify the nature of binders and to distinguish between natural hydraulic mortars, natural cements and Portland cements. FTIR, XRD analysis and SEM observation have been employed to identify gehlenite and di-calcium silicates as characteristic hydraulic phases of natural hydraulic lime, whereas a large amount of C_3S, is generally associated to the higher burning temperature occurring during cement's production.

The restoration materials used in the conservation field must fulfil special requirements in order to be compatible with the original ones [8]. As a matter of fact, the current lack of information about commercial ready-mixed mortars' composition and durability at work may complicate the selection of the most suitable restoration materials based upon the compatibility evaluation.

The present work proposes a methodology for the characterization of ready-mixed mortars for the conservation of the architectural heritage and reports on the results of its application on four widely diffused commercial products.

Compositional, morphological and mechanical features are discussed and compared.

2 Experimental

2.1 Products and supplier

Four commercial ready-mixed mortars designed for restoration purpose, all classified as cement-free, with a minimum content of soluble salts and based on NHL – natural hydraulic lime binder were selected. The sample materials were named MA, MB, MC and MD without any reference to the manufacturer.

2.2 Specimens preparation

Mortars were prepared according to the indication reported in the technical data sheets. Casting was performed following two different procedures: a) traditional casting in 4x4x16 cm demountable steel mould, b) porous substrate casting. The porous substrate for the latter type of specimen was provided by fire-clayed brick, type "San Marco". Bricks underwent 12-hours water imbibition followed by the removal of the superficial liquid water layer before mortars' application. Specimens were let to harden for 48 hours at 20°C – 90% RH. Specimens prepared following traditional casting were then removed from the mould, while specimens prepared on porous substrate were stored together with their brick substrates. The curing conditions were 20°C – 90% RH for 60 days for specimens on porous substrate and for 28 days and 60 days for traditional prismatic specimens.

2.3 Analytical techniques

The initial composition of the mortar mixture was analysed by X-ray diffraction and IR spectroscopy on anhydrous powder samples. X-ray diffraction was carried out by means of a Philips PW1830 instrument using Cu Kα radiation (λ = 1.54058 Å). FTIR spectroscopy was carried out on a Thermo Nicolet 6700 instrument. Fine grinded samples were analysed after dispersion in KBr pellets (KBr FTIR grade by Sigma-Aldrich).

The mechanical behaviour of hardened mortars was studied on traditional 4x4x16 cm specimens while the observation of the microstructural features was made on specimens prepared on porous substrate.

Compressive strength was evaluated after a curing time of 28 days and 60 days, according to standard protocol [9]. Five specimens of each mortar were tested and the reported results are average values of 5 different measurements.

Polarization-and-fluorescence microscopy was performed on thin sections. Mortars' fragments were vacuum impregnated with a UV-fluorescent resin and observed with a Zeiss Axioplan microscope. Further microstructural characterization was performed by scanning electron microscopy on freshly fractured samples. An Environmental Scanning Electron Microscope (ESEM)

Zeiss EVO 50 EP equipped with an Oxford INCA 200 – Pentafet LZ4 spectrometer was used in secondary electrons mode.

3 Results

3.1 XRD analysis

The diffractograms reported in fig. 1 highlight the abundance variability of the crystalline phases of the four analysed mixes.

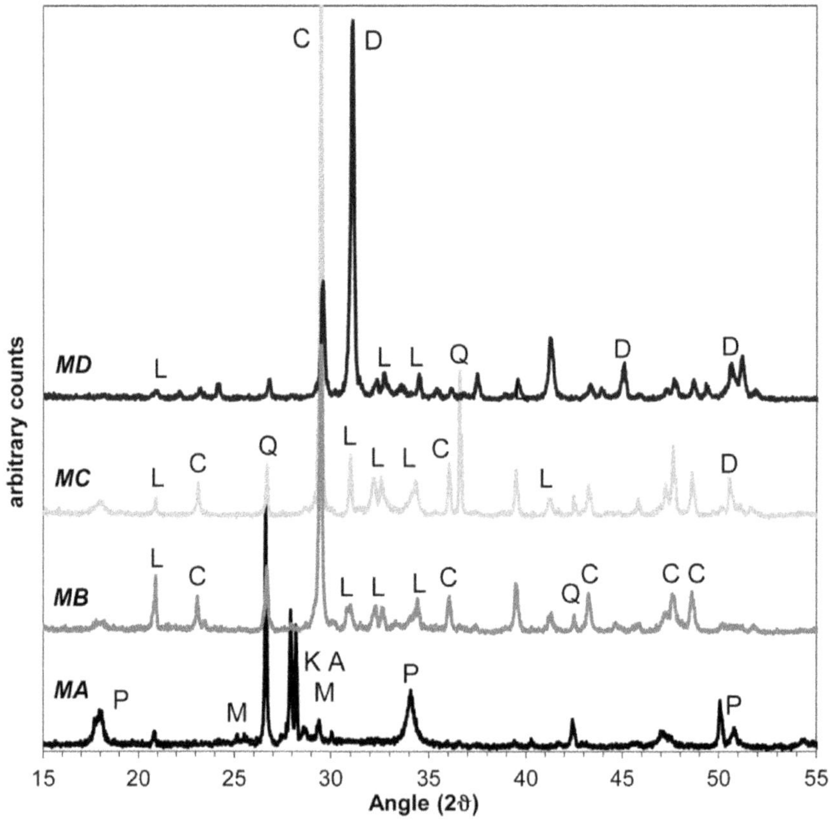

Figure 1: X-ray diffractograms of anhydrous powder samples of the four studied mortars (A = albite; C = calcite; D = dolomite; K = kyanite; L = larnite; M = muscovite; P = portlandite; Q = quartz).

The XRD results show significant differences in the binder and aggregate composition of the four examined mortars. As far as the binder is concerned, mortar MA is the only one showing the characteristic peaks of portlandite, $Ca(OH)_2$. A certain amount of portlandite is expected to be found in NHL, as unreacted calcium oxide can derive from calcination of the binder and it is transformed into hydroxide after slaking. In all other cases, results only show the presence of larnite (Ca_2SiO_4, i.e. an analogue of C_2S) in the binder composition, thus confirming the hydraulic behaviour of the mortar.

Calcite and quartz are present in all samples with different ratios and they belong to the aggregate fractions. The calcite content is particularly high in the case of mortar MB. MA and MB also contain micas and plagioclase. MC aggregate is partially composed of dolomite, which can be observed as the most abundant crystalline phase in MD. Therefore the aggregate phases of MA and MB can be defined as rather heterogeneous, made of quartz-siliceous and carbonatic compound, whereas those of MC and MD are mostly carbonatic.

3.2 FTIR analysis

The results of FTIR characterization of anhydrous commercial materials are reported in fig. 2.

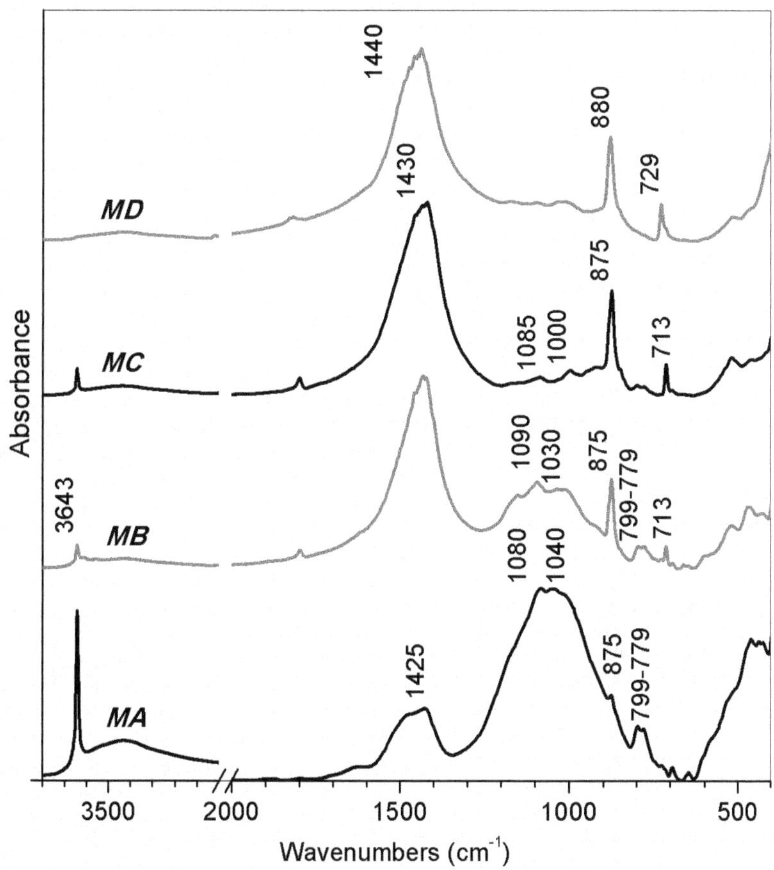

Figure 2: FTIR spectra of anhydrous mortars' samples in the spectral range 4000–400 cm^{-1}.

MC and MD spectra are quite similar. In both cases the absorption peaks indicate that a large amount of carbonates are present as calcite (peaks at 1430, 875, 713 cm^{-1}) in MC sample, and as dolomite (1440, 880, 729 cm^{-1}) in MD sample. MC also contains a significant amount of free portlandite, with a sharp peak at 3643 cm^{-1}. Si-O and Si-O-Si bonds vibrational modes are detectable in the range of frequencies between 1100–750 [7] and are associated to both the quartz-siliceous aggregates and to the hydraulic compounds of the binders: quartz around 1080 cm^{-1} with a characteristic doublet at 797 and 779 cm^{-1}; calcium silicates at 1000–1010 cm^{-1} and 930 cm^{-1}; pozzolan or other poorly crystalline siliceous phases giving a broad peak around 1035–1030 cm^{-1}. Peaks related to the presence of quartz can be detected in both samples, while those related to the presence of calcium silicates are more evident in sample MC.

MB sample contains calcite and portlandite, similar to the samples discussed above. However, this mortar show a high absorption in the region of Si-O and Si-O-Si region, where quartz (1093 cm^{-1}), calcium silicates, amorphous compounds and pozzolan (1030 cm^{-1}) peaks are detected.

Finally, MA sample shows a quite different vibrational spectrum in the mid-IR region. In particular, the C-O/Si-O ratio is reversed in this case respect to the previous ones, with quartz/siliceous peaks prevailing over carbonates. This is due to the fact that the aggregate fraction is rich in quartz (1082, 797, 779 cm^{-1}) and siliceous phases as already pointed out by XRD, whilst calcite (1425, 876 cm^{-1}) is present in lower amount. Moreover, the broad overlapped absorption peak around 1040 cm^{-1} is also related to the presence of hydraulic compounds, namely calcium silicates. The sharp and very intense peak at 3643 cm^{-1} indicates the abundant presence of portlandite.

3.3 Petrography

Thin sections observations were carried out on specimens prepared on brick surfaces in order to properly evaluate the influence of a porous substrate on the microstructure of the hardened mortars. The results are presented according to the different commercial materials.

Mortar MA is composed of a hydraulic lime binder with irregular porosity. No cracks or fractures are present (fig. 3): the mortar's matrix shows a quite compact microstructure with a fine grained binder.

Aggregates are poorly sorted and vary in size up to 2 mm. Quartz is confirmed to be the most common aggregate, with rounded shape. Limestone, calcite, feldspar, plagioclase and few amphibolites are also present. Limestone is generally present as coarse, rounded shape aggregate while angular calcite grains are smaller in size. The very high amount of fine grained calcite crystals indicates that calcite has been used as filler. A few small grains of pozzolan are also present.

The hydraulic phases of the binder are uniformly dispersed within the binder matrix. Few grains of not reacted C_2S are detected [10]. A carbonation front proceeds from the external borders into the bulk, with some isolated carbonated patches at deeper levels, along pores.

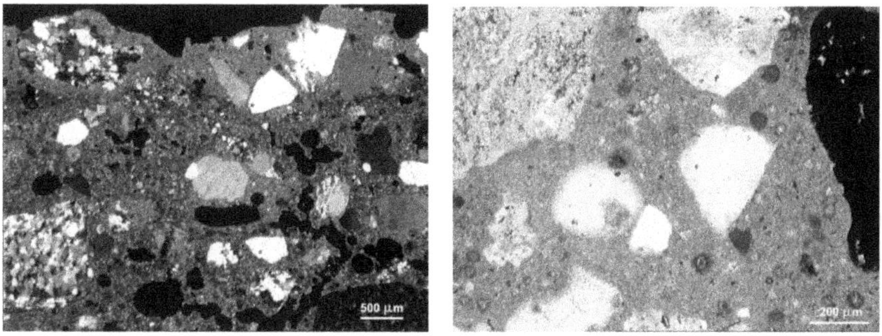

Figure 3: Mortar MA: carbonation front and irregular porosity (left, CPL); dark coloured not reacted C_2S particles dispersed in the carbonated binder matrix (right, PPL).

Mortar MB has a completely different binder structure respect to the others (fig. 4). Porosity is irregular and generally lower than observed in mortar MA. Pores and voids of different shape are connected by a network of micro-cracks. Cracks occur both in the binder matrix and along the binder-aggregate interface.

Aggregates are heterogeneous; quartz and calcite are the main constituents, together with phyllosilicates and few amphibolites. Calcite is present as rounded limestone particles and as crushed angular fragments. The mortar has a very dense microstructure. No traces of hydraulic compounds, such as C_2S grains, are detected. Few grains of not reacted pozzolan are observed. The coarse grained binder is very rich in slag fragments, which appear as transparent angular vitreous grains and are homogeneously dispersed in the mortar. Carbonation is limited to a thin region at the margin of the sample and along pores walls. Some fractures cross the binder's matrix, as observed in mortar MA.

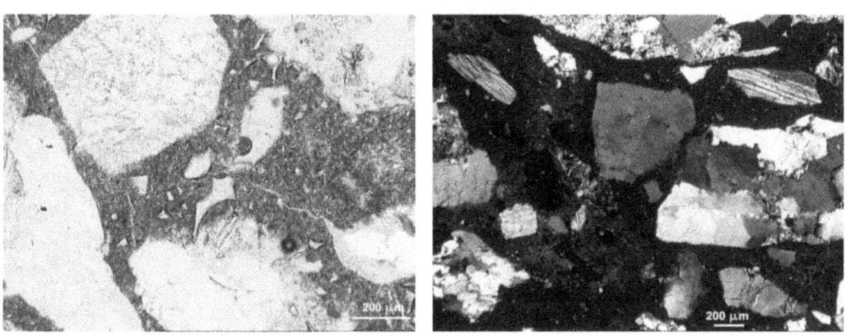

Figure 4: Mortar MB: angular slag fragments and micro-cracks in the binder matrix (left, PPL); aggregates dispersed in a dark binder matrix (right, CPL).

Mortar MC is composed of a lime binder with an apparently high porosity due to the presence of abundant air voids (fig. 5). This can most probably derive from the introduction of an air-entraining agent in the mortar mix. Some de-bonding between mortar and aggregate occurred, but cracks are generally absent in the mortar's matrix.

Aggregates are poorly sorted and vary in size up to 1.5 mm. Carbonatic aggregates are dominant. No quartz aggregate can be observed.

The binder is coarse grained, with not reacted C_2S particles heterogeneously distributed in the matrix. Very few C_3S grains surrounded by a clear hydration rim occur. Some pozzolan particles are present.

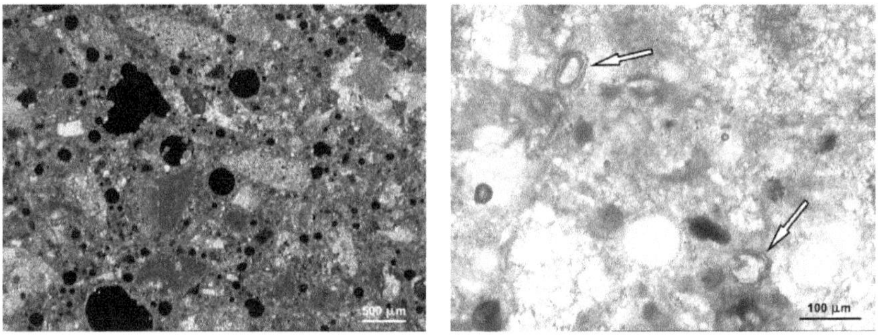

Figure 5: Mortar MC: calcite and limestone aggregate and abundant air voids (left, CPL); arrows indicate partially hydrated C3S particles in the binder matrix (right, PPL).

Mortar MD has a hydraulic lime based binder with irregular porosity (fig. 6). Small air voids occur together with large irregular ones. The mortar shows no micro-cracks.

Aggregates are poorly sorted and vary in size up to 2 mm. Limestone, crushed angular calcite and quartz, are the main aggregates; minor feldspar is also present. Limestone occurs as coarse rounded grains while the smaller aggregate fraction is composed of quartz and crushed calcite. No pozzolan grains are present.

Figure 6: Mortar MD: limestone and quartz aggregates and non-homogeneous distribution of C2S particles (left, PPL); carbonation front along the margin (right, CPL).

The distribution of hydraulic phases within the binder matrix is not homogeneous and several darker zones due to high C_2S concentration are present. A thin carbonation front proceeds from the external margins of the sample.

3.4 ESEM observation

ESEM observation of the hardened mortar samples allows the identification of the newly formed structure related to the hydration process of the binder. In all examined samples hydrated calcium silicate phases (CSH) are clearly visible, and they have a variable morphology (fig. 7).

Mortar MA shows a diffuse fibrous CSH network covering most of the aggregate grains, with occasional isolated clusters of needle shaped CSH [11]. Within the CSH network, some quartz grains can be observed together with few non reacted pozzolan particles, appearing as well defined globular grains with a smooth surface.

Mortar MB shows a denser CSH structure as a result of the hydration of the hydraulic compounds together with the slag fragments (fig. 7a). These latter can hardly be identified as they are almost completely covered by the hydrated phases. CSH develops globular cluster connected by numerous and particularly elongated crystals. No pozzolan can be observed.

Sample MC (fig. 7b) and sample MD have a similar microstructure. Globular CSH are crossed by a number of acicular, needle shaped crystals. Some isolated clusters of hydrated particles occur. A significant amount of portlandite crystals, occurring as typical hexagonal plates, are embedded within the binder's matrix.

3.5 Mechanical test

As might be expected for NHL mortars, compressive strength results of the mortars (fig. 8) show a considerable increase as the curing time proceeds [12].
The average compressive strength values after 28 days curing vary considerably. MC and MD are based respectively on NHL 3.5 and NHL 5 binder [1], thus their strength value is in accordance with the class of resistance they belong to only in the case of MC. On the contrary, MD has a 28 days resistance significantly lower than the 5 MPa limit expected for its class. Similar consideration cannot be extended to MA and MB because of the lack of information about the classes of binder used in mixes. MD shows indeed the lowest compressive strength among all the tested mortars after 28 days curing, while both MA and MB have values higher than 9 MPa.

As the hydration of the hydraulic compounds takes place, the compressive strength resistance increases in all cases, except for MA which shows a slight decrease respect to the 28 days value. It can be observed that the significant initial differences between the mortars' resistance tend to be reduced during curing. MA and MB have a rather similar mechanical behaviour between 9.5 and 10 MPa. The final compressive strength of MC is slightly lower, around 9 MPa, and MD is confirmed to be the less resistant mortar.

Figure 7: ESEM images of the microstructure of a) mortar MB, showing a dense CSH structure with elongated crystals; and b) mortar MC, showing hexagonal portlandite crystals (1) and CSH particles (2).

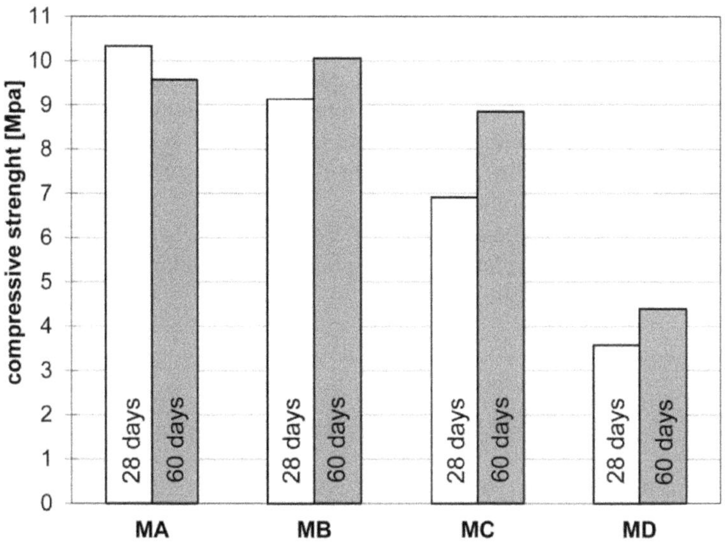

Figure 8: Average compressive strength of the mortars tested after a curing time of 28 days (left row) and 60 days (right row).

4 Conclusions

The investigation of four commercial ready-mixed mortars allows the discrimination of distinctive compositional and microstructural features of the tested products. MA has a portlandite-rich binder with some pozzolan and limestone filler. During hardening a diffused CSH network homogeneously covers the aggregates and may explain the high compressive strength after 60 days curing.

MB is, unexpectedly, a slag-lime mortar. The hydraulic behavior of the mortar is given by the hydration of the slag and by the presence of pozzolan. The hardened mortar has a compact structure and the 60 days compressive strength is comparable with that of MA.

At first glance, analyses show that MC and MD have a rather similar composition. They both have a high amount of calcareous aggregate (mainly calcite in MC and dolomite in MD). However, the microstructure of MC differs in one aspect, i.e. the air void content, which is probably due to the presence of an air-entraining agent in the mix. MC is also the only sample clearly showing some partially hydrated C_3S grains. C_3S is generally uncommon in NHL mortars and may derive from a higher burning temperature respect to the other products. As a consequence, MC and MD show different compressive strengths. The high strength of MC may be related to the presence of C_3S in the binder and it seems not be affected by the apparently high porosity of the mortar, while the lower one of MD can be influenced by the predominant dolomitic aggregate.

The analytical methodology adopted for the characterization of the ready-mixed mortar samples proved to be useful to discriminate the products. Though they are supposed to be comparable NHL mortars, they show different compositional features, microstructures, presence of additives and mechanical behaviour. These are essential to evaluate compatibility with historic materials and should be stated on the technical data sheets, or they have to be thoroughly determined prior to the application.

Additional measurements, such as pores structure, pores size distribution and hydraulic Si amount could be useful for a better understanding of these products and to evaluate their durability at work.

References

[1] UNI EN 459-1.Building Lime – Definitions, specification and conformity criteria, 2010.

[2] Allen, G., Allen, J., Elton, N., Farey, M., Holmes, S., Livesey, P. & Radonjic, M., *Hydraulic Lime Mortar for Stone, Brick and Block Masonry*, Donhead: Shaftesbury, pp. 3–10, 2003.

[3] Groot, C., Ashall, G. & Hughes, J., Characterisation of old mortars with respect to their repair. Final report of RILEM TC 167-COM, 2004.

[4] RILEM Technical Committee, Repair mortars for historic masonry. Testing of hardened mortars, a process of questioning and interpreting. *Materials and Structures*, 42, pp. 853–865, 2009.

[5] Maravelaki-Kalaitzaki, P., Bakolas, A., Karatasios, I., Kilikoglou, V. Hydraulic lime mortars for the restoration of historic masonry in Crete. *Cement and Concrete Research*, 35(8), pp. 1577–1586, 2005.

[6] Callebaut, K., Elsen, J., Van Balen, K. & Viaene, W., Nineteenth century hydraulic restoration mortars in the Saint Michael's Church (Leuven, Belgium) Natural hydraulic lime or cement? *Cement and Concrete Research*, 31, pp. 397–403, 2001.

[7] Varas, M.J., Alvarez de Buergo, M. & Fort, R., Natural cement as the precursor of Portland cement. *Cement and Concrete Research*, 35, pp. 2055–2065, 2005.

[8] Rodrigues, J.D., Grossi, A., Indicators and ratings for the compatibility assessment of conservation actions. *Journal of Cultural Heritage*, 8(1), pp. 32–43, 2007.

[9] UNI EN 1015-11.Methods of test for mortar for masonry: Determination of flexural and compressive strength of hardened mortar, 2007.

[10] Elsen, J., Microscopy of historic mortars. A review. *Cement and Concrete Research*, 36(8), pp. 1416–1424, 2006.

[11] Diamond, S. & Kjellsen, K.O., Resolution of fine fibrous C-S-H in backscattered SEM examination. *Cement & Concrete Composites*, 28, pp. 130–132, 2006.

[12] Lanas, J., Sirera, R. & Alvarez, J.I., Study of the mechanical behaviour of masonry repair lime-based mortars cured and exposed under different conditions. *Cement and Concrete Research*, 36, pp. 961–970, 2006.

A procedure to assess the suitability of plaster to protect vernacular earthen architecture

E. Hamard[1], J. C. Morel[1], F. Salgado[2], A. Marcom[3] & N. Meunier[4]
[1]*Ecole Nationale des Travaux Publics de l'Etat – ENTPE, France*
[2]*Pontifícia Universidade Católica do Rio de Janeiro – PUC-Rio, Brasil*
[3]*Inventerre, Saint Pierre de Lages, France*
[4]*N. Meunier, Chambles, France*

Abstract

As part of a working definition of a new code of practice, a methodology is being defined to determine the suitability of plasters manufactured on-site to protect the earthen walls of vernacular architecture buildings. Given the diversity of raw earth construction methods, ranging from massive earth to stone masonry with earth mortars, and the variability of the materials used, two types of on-site testing were proposed: a shrinkage test followed by a shear test. These tests, as well as additional tests, were performed on an earthen wall using two different earth plasters. This first testing campaign has permitted the validation of the on-site shrinkage test but not of the on-site shear test. A future testing campaign will explore the influence of various parameters on the shear test.
Keywords: earth plaster, earthen vernacular architecture, durability, on-site coating testing.

1 Introduction

Historic earthen buildings represent an important, though difficult to quantify, part of the human habitat worldwide. In France, according to Michel and Poudru [1], in 1987 there were approximately one million earth houses (rammed earth, adobe and cob) with an average age of 100 years. All of them are more than 50 years old. This heritage ought to be preserved. One of the major issues is the repair of their protective plasters. Unfortunately, old construction techniques fell into disuse in the West and therefore one can no longer rely on the vast majority of today masons. Industrial ready-to-use coatings are unsuitable and even harmful to earth constructions [2, 3]. As part of a working definition of a new code of practice for plaster implemented on earthen walls, a research campaign was conducted to validate on-site tests to select suitable plaster formulations. This study focuses on the earth plasters' mechanical behaviour.

Earth plasters have two advantages: they are permeable to water vapour and their mechanical behaviour is close to earthen walls, which make them more compatible than industrial coatings that are waterproof and too stiff [2, 3]. Indeed, the mechanical compatibility can be estimated taking into account the difference between the Young's modulus of the wall and the plaster. If this difference is too high, the stress changes induced by overloads, by moisture and by temperature variations, generate a differential strains between the plaster and the wall. Those two elements are linked, leading to damages in the plaster or in the earthen wall. Earthen wall Young Modulus (ranging from 0.5 to 1 GPa for rammed earth [4, 5]) are much lower than concrete modulus (ranging from 15 to 50 GPa). That is why earth plasters (close to earth mortar and adobe with a Young modulus ranging from 0.5 to 2.5 GPa [6]) and lime/sand plasters [7] are *a priori* well adapted to earthen walls.

Earth plasters are made from clayey soils called earths, extracted near the construction site, to which a certain amount of sand is added. Numerous admixtures can be added to these two bases, as masons used to do for centuries, to improve their behaviour, for example: vegetable and animal fibres, soap, milk, fresh cow dung, red wine [8–11]. Unlike standardized building materials, materials entering in the composition of earth plasters have a considerable variability. It is not possible to propose a formulation assessment adapted to all sites. This is why, inspired by tests already carried out by the masons, field tests are proposed to validate plasters' formulations, whatever the nature of materials used. This approach is designed to allow the choice of materials at the discretion of the mason, thus promoting the use of local materials and respecting local construction cultures.

An earth plaster of acceptable mechanical quality is a plaster which, after shrinkage, has no cracks that could allow water penetration into the wall and has a sufficient bond with its wall. Here, two on-site tests are proposed to validate the plaster formulations: a shrinkage test and a shear test. An initial testing campaign was conducted to validate these two tests. The proposed tests were applied to plasters made from two different earths. Additional laboratory tests were conducted to measure the shrinkage, the tensile strength as well as the water content of these plasters.

2 Experimental program

2.1 Earthen wall preparation

For testing, an earthen wall was used, allocated in a farm garden belonging to the *Grand Parc de Miribel Jonage* (Lyon, France). This wall is composed of a base of slag concrete that underpins an earth rising by about 500 mm. Before testing, the wall was brushed and repeatedly wetted with a garden sprayer. A scratch coat consisting of 1 volume of NHL 3.5 hydraulic lime and two volumes of Hostun sand [12] was thrown on the earthen wall. Earth plasters are applied directly on this scratch coat.

2.2 Testing campaign

Earth plasters were prepared from earth (clays, silts and sand) with variable proportion of sand and water added. Clays play the role of binder and are responsible for the shrinkage of the plaster. Silts and sands form the granular skeleton that gives the mechanical structure of the plaster. Most of the time, there is not enough sand in the natural earth and this is why it had to be added.

A good earth plaster formulation must contain enough clay to bind the entire granular skeleton and avoid its erosion, but not too much to limit its shrinkage. Therefore it is usual to seek the earth-to-sand ratio by volume or by mass which will generate no harmful shrinkage to plaster and will allow a sufficient sand grains covering. In this study, shrinkage is considered harmful if it causes deep cracks or a bow out of the plaster; cracking is not considered injurious.

The shrinkage test requires a sample for each formulation created in two layers of 250 mm by 250 mm brown coat, 20 mm thick. After drying, when shrinkage is completed, the presence or absence of shrinkage cracks in the samples was noted. Only samples with no shrinkage cracks were validated.

A second important point to validate in earth plaster formulations is their bond on the wall. This bond depends on the nature of the wall (sources of materials, material implementation and possible heterogeneities), on the wall hydric state and on the nature of the plaster. Since the tests were performed in the closest possible on-site conditions and on different parts of the wall, these parameters need not be considered.

Two samples for each formulation, 50 mm × 40 mm and 20 mm thick, were shaped in a single layer for the shear test. The 250 mm × 250 mm samples were

Figure 1: Load device.

also used in the shrinkage test. They were cut in order to get 4 samples of 50 mm × 40 mm and 20 mm thick. Thus, 6 samples were obtained per formulation: 2 "shaped" and 4 "cut". Samples were loaded up to failure with a load device (fig. 1), by increments of 0.5 kg every 60 s for T17.5, T12, T9 and T6 samples (table 1 and 2) and then, for reasons of precision, loaded by increments of 0.25 kg every 30 s for the other samples.

Knowing the loads and the area of the sample (S, mm^2) it is possible to calculate the shear rate (τ, Nmm^{-2}):

$$\tau = \frac{m_f g}{S}$$

with $g = 9.81\ ms^{-2}$.

Additional laboratory tests were carried out to better characterize the earth plaster:

- Measurement of the plaster's longitudinal shrinkage for samples of 40 mm × 40 mm × 160 mm;
- Three points bending test on plaster samples of 40 mm × 40 mm × 160 mm.

The maximum force reached (F, N) was recorded and the bending strength (σ, Nmm^{-2}) was calculated:

$$\sigma = \pm \frac{1{,}5Fd}{wh^2}$$

where:

- d (mm) is the distance between lower supports points,
- w (mm) is the width of the specimen,
- h (mm) is the height of the samples.

The bending strength is equal to the tensile strength, since tensile failure occurs before compression in this pure bending test.

2.3 Sample production

Two earths were used in the tests: Tassin and Rochechinard earths [6] (table 1). Four different clay contents were considered: 17.5%, 12%, 9% and 6% in dry weight by adding Hostun sand [12]. The Rochechinard earth plasters with 17,5% and 12% were also tested with 0.5% admixture by dry weight, Sisal fibres [13] and hemp chaff aggregate [14] (table 2). Finally a lime-sand plaster, composed of 1 volume of lime (0.5% calcic lime and 0.5% natural hydraulic lime NHL 3.5) and 3 volumes of Hostun sand, was also tested.

Table 1: Earths composition (MBV = Methylen Blue Value).

	Clays < 2 µm	**2 µm < silts < 20 µm**	**20 µm < sands< 2000 µm**	**MBV**
Tassin	17.5%	12.5%	70%	1.4
Rochechinard	25.5%	32.5%	42%	2.5

Table 2: Earth plasters formulations.

Earth	Tassin				Rochechinard				Rochechinard with admixtures			
Sample name	T17.5	T12	T9	T6	R17.5	R12	R9	R6	R17.5 + S	R12 + S	R17.5 + H	R12 + H
Clay (%)	17.5	12	9	6	17.5	12	9	6	17.5	12	17.5	12
Earth (kg)	5	3.43	2.57	1.71	3.43	2.35	1.76	1.18	3.43	2.35	3.43	2.35
Sand (kg)	0	1.57	2.43	3.29	1.57	2.65	3.24	3.82	1.57	2.65	1.57	2.65
Hemp chaff (g)	-	-	-	-	-	-	-	-	-	-	25	25
Sisal (g)	-	-	-	-	-	-	-	-	25	25	-	-

Before the test, earth and sand are placed in an oven to dry. Then earth-sand mixtures were fabricated in the right proportions in the laboratory and bagged to keep them dry. On-site, the water amount considered necessary to obtain a good workability of the mortar was added. This amount of water was measured; this allowed the calculation of their water content.

3 Results and discussion

3.1 Shrinkage tests

The results of the 250 mm × 250 mm squares shrinkage test are presented in fig. 2. Tassin and Rochechinard 17.5% clay plasters fell from the wall; those with 12% clay cracked and 9% and 6% clay did not crack. It was noted that the plasters with 6% clay were very sandy, which make them more difficult to work and relatively powdery after drying. Optimal clay content in this case is around 9%. Clay content has a clear influence on earth plaster cracking. Tassin and Rochechinard earths presented a similar workability and shrinkage for the same clay content. This observation is consistent with the Methylen Blue Value of those earths which are of the same order (table 1).

The addition of Sisal fibres in the Rochechinard earth with 17.5% and 12% clay results in great improvement in the workability of the mortar. It is the same for cracking, since the 17.5% clay plaster without fibre, which fell from the wall, shows no cracks when 25 g of Sisal were added, even if it tends to bowed out. The 12% clay plaster with Sisal fibres would be validated unlike its counterpart without fibres.

Rochechinard earth plus 25 g hemp chaff aggregate has a workability identical to the one without fibres. The addition of 25 g of hemps chaff does not therefore improve the workability of the mortar. The shrinkage causes a bow out and some cracks on these earth plasters (R17.5 + H and R12 + H). Even if cracking is less important for plasters with hemp chaff than for those with earth, none of those plasters could be validated.

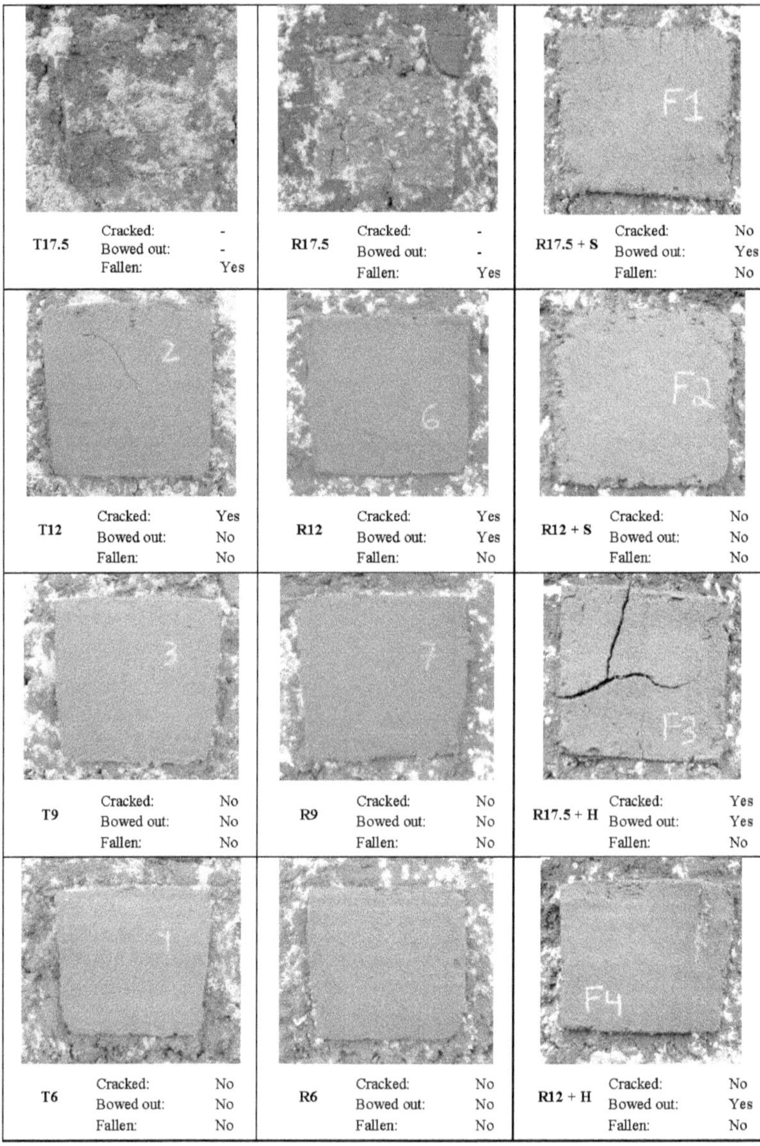

Figure 2: Results of the shrinkage test.

3.2 Shear tests

Since it was not possible to shape or cut the samples at the exact sizes, the actual dimensions were measured and calculations were made with the real values. Some 250 × 250 mm samples could not be cut for the shear test, either because they fell before the test (T17.5 and R17.5), or because they were too bowed out and too cracked to be tested (R17.5 + H). Some samples badly supported cutting and could not be tested (one of the R17.5 + S, and two of the R12 + H samples, as well as all the lime-sand samples). The shear test results showed (fig. 3):

- a greater strength for shaped samples compared to cut samples, which were cut with an angle grinder that damaged earth plaster and support due to vibration. Shear strength values are smaller and show greater spread for cut samples than for shaped ones. Therefore cutting of samples should not be allowed;
- an optimum shear strength for a content of clay ranging between 6% and 8%;
- a greater strength of lime-sand plasters samples in comparison to those made from earth;
- a large dispersion of the results that can be attributed to the heterogeneity of the wall.

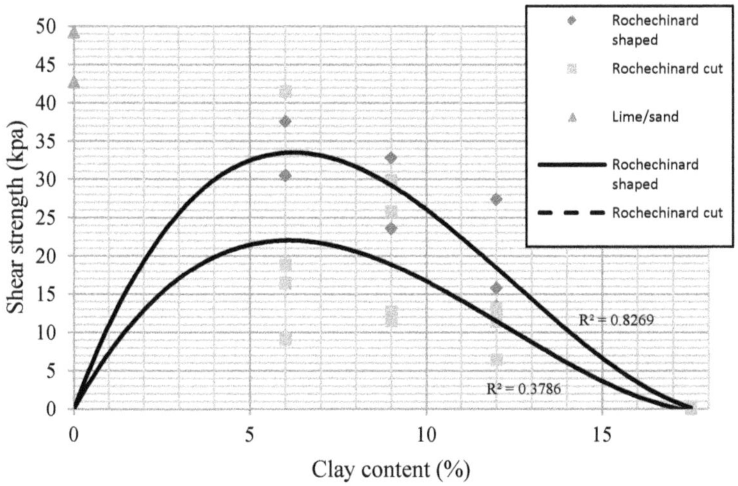

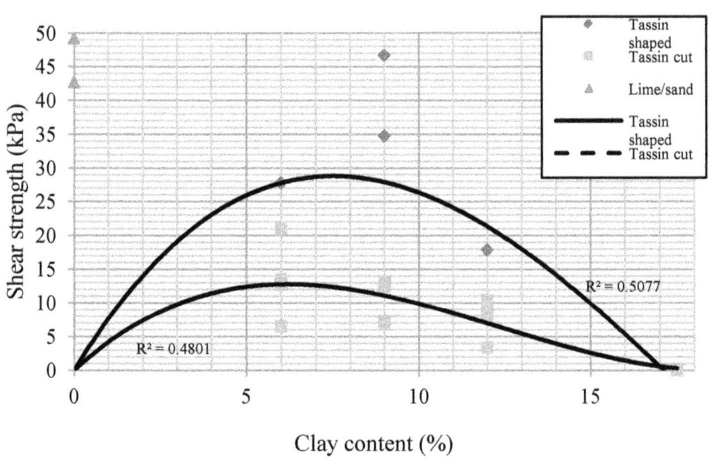

Figure 3: Influence of clay content on shear strength for Rochechinard and Tassin earths.

3.3 Bending strength

Three-point bending strength test results (fig. 4) revealed that:

- the bending strength increases with the increase of clay content;
- the lowest bending strength of the investigation was that of the lime-sand plasters. This can be explained by the use of the very fine Hostun sand (maximum diameter of 0.8 mm) that did not structured enough the plaster. It has probably happened also because the time lapse between manufacture and test (1 month) was not sufficient to enable a good setting of the calcic lime.

If increasing clay content in the earth plaster improves its bending strength (fig. 4), the shrinkage, which results from clay content increase, limits and then cancels this effect (fig. 2). This can explain the existence of an optimum shear strength of earth plasters (fig. 3).

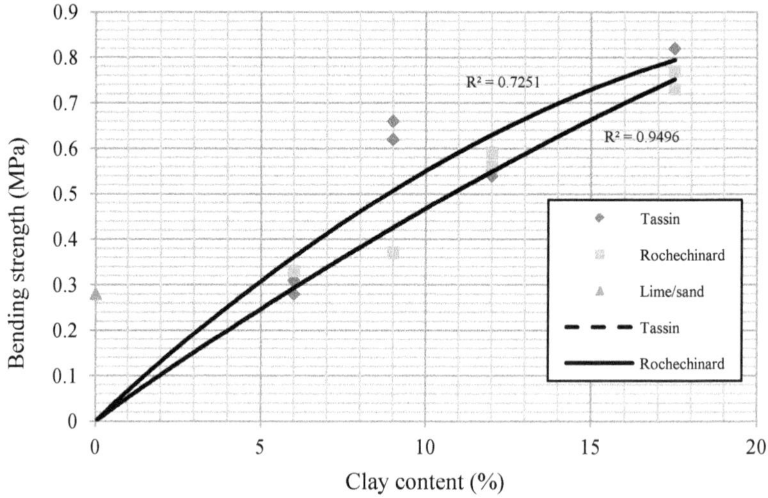

Figure 4: Influence of clay content on three point bending strength of earth plasters.

4 Conclusion

A significant current problem is the preservation of earthen vernacular buildings, which bear testimony of local constructive cultures and values of sustainable development. Non industrial materials used with raw earth require a suitable scientific approach.

The first test campaign has been very satisfactory for the on-site shrinkage test. Its experimental protocol can be validated. For the on-site shear test, cut samples were weakened and they cannot be considered in the experimental campaign. A future testing campaign will explore more systematically the on-site shear tests. This might allow us to better understand the influence of the different parameters that control shear resistance of the plaster on the wall and to propose well adapted on-site tests to assess the suitability of plasters to protect vernacular earthen architecture.

Acknowledgements

The authors wish to gratefully acknowledge Prof. Ghavami for having helped Fernanda Salgado do her Master course in France, the whole team who worked to realise the code of practice for their contribution and the *Grand Parc de Miribel Jonage* for providing us the *Allivoz* farm fence wall to allow us to perform our tests.

References

[1] Michel, P. & Poudru, F., Le patrimoine construit en terre en France, actes du colloque international sur le patrimoine européen construit en terre et sa réhabilitation, ENTPE, pp. 529–549, 1987.

[2] Lanas, J. & Alavarez, J.L., Masonry repair lime based mortars: factors affecting the mechanical behaviour. *Cement and Concrete Research*, **33** pp. 1867–1876, 2003.

[3] Walker, P., Keable, R., Martin, J. & Maniatidis, V., Rammed earth – design and construction guidelines, BRE Bookshop, 2005.

[4] Bui, B., Morel, J.C., Hans, S. & Meunier, N., Compression behaviour of non-industrial materials in civil engineering by three scale experiments: the case of rammed earth. *Material and Structures*, **42**, pp. 1101–1116, 2009.

[5] Minke, G., *Earth Construction Handbook, the Building Materials Earth in Modern Architecture*, WIT Press, Southampton, UK, 2009.

[6] Alves de Azeredo, G., Mise au point de procédures d'essais mécaniques sur mortier de terre: application à l'étude de leur rhéologie, PH.D. thesis, INSA Lyon, France, 2005.

[7] Arandigoyen, M. & Alvarez, J.L., Pore structure and mechanical properties of cement-lime mortars. *Cement and Concrete Research*, **37(5)** pp. 767–775, 2007.

[8] Yetgin, S., Cavdar, Ö. & Cavdar, A., The effect of the fiber content on the mechanic properties of the adobes. *Construction and Building Materials*, **22(3)**, pp. 222–227, 2008.

[9] Achenza, M. & Fenu, L., On earth stabilization with natural polymers for earth masonry construction. *Material and Structures*, **39**, pp. 21–27, 2006.

[10] Quagliarini E. & Lenci, S., The influence of natural stabilizers and natural fibres on the mechanical properties of ancient Roman adobe bricks. *Journal of Cultural Heritage*, **11(3)**, pp. 309–314, 2010.

[11] Galán-Marín, C., Rivera-Gómez, C. & Petric, J., Clay-based composite stabilized with natural polymer and fibre. *Construction and Building Materials*, **24**, pp. 1462–1468, 2010.

[12] Flavigny, E., Desrues, J. & Palayer, B., Note technique – le sable d'hostun in Revue Française. *Géotechnique*, **53**, pp. 67–70, 1990.
[13] Ghavami, K., Toledo Filho, R.D. & Barbosa, N.P., Behaviour of composite soil reinforced with natural fibres. *Cement and Concrete Composites*, **21**(**1**), pp. 39–48, 1999.
[14] Samri, D., Analyse physique et caractérisation hygrothermique des matériaux de construction: approche expérimentale et modélisation numérique, Ph.D. thesis, INSA Lyon, France, 2008.

Hygrothermal performance of innovative renovation renders used for different types of historical masonry

J. Kočí[1], J. Maděra[1], P. Rovnaníková[2] & R. Černý[1]
[1]*Czech Technical University in Prague, Czech Republic*
[2]*Brno University of Technology, Czech Republic*

Abstract

A combined experimental-computational approach is applied for the estimation of service life of innovative renovation renders. In the experimental part, the durability of two commercial renovation renders is determined in terms of their frost resistance, together with two newly developed renovation plasters that are supposed to be produced commercially in the near future. The computational part is aimed at the investigation of freeze-thaw cycles that may occur in surface layers of the plaster during its life time period. To achieve this, a diffusion type model of coupled heat and moisture transport is used. As a load-bearing structure several materials which are characteristic for historical buildings are chosen, namely ceramic brick, sandstone and arenaceous marl.

Keywords: service life, renovation renders, heat transport, moisture transport, experimental analysis, frost resistance.

1 Introduction

In the current practice of historical buildings renovation, the service life of new surface layers is estimated mostly on the basis of experience, because too few parameters are known for reliable durability estimate. Usually the durability is given by the producer of the renovation render, who provides a warranty period only in the case that the technological process is followed precisely. The service life of a new surface layer made of renovation render is usually declared between 10 and 15 years. Unfortunately, the real service life is much lower if we take into account climatic conditions, location of the building, local building environment and last but not least, the material composition. As has been proven [1–4], a proper material configuration of the building envelope may significantly extend its service life. Combination of improperly chosen materials can lead to damage caused by moisture transport through the interface between materials [5].

In this paper, a method, not frequently used until now, is applied for the solution of the problem of damage of surface layers of historical buildings. The method consists of a combination of computational and experimental approaches. In the experimental part, the durability of several innovative renders developed under laboratory condition on a lime-pozzolana basis is determined in terms of their frost resistance. In the computational part, the number of frost cycles in a real structure is identified using a computational analysis. In addition, the computer code involves the effect of hysteresis of moisture transport and storage parameters; this brings the results of the computations nearer to reality. With the knowledge of the yearly number of freeze-thaw cycles in the surface layers, the service life of renovation renders can be determined by comparison with laboratory experiments.

Prediction of moisture and temperature fields in the building envelope parts of historical buildings over a sufficiently long time makes it possible not only for a reliable assessment of durability of surface layers on the basis of a sufficient amount of input parameters, but also for a design of the time schedule of repeated repairs. It can also offer to the supervisory authorities alternative solutions of surface layers taking into account their durability and financial demand.

2 Studied building envelopes

The studied building envelope is presented in fig. 1. It consisted of a load-bearing structure and single layered renovation plaster on the exterior and interior side.

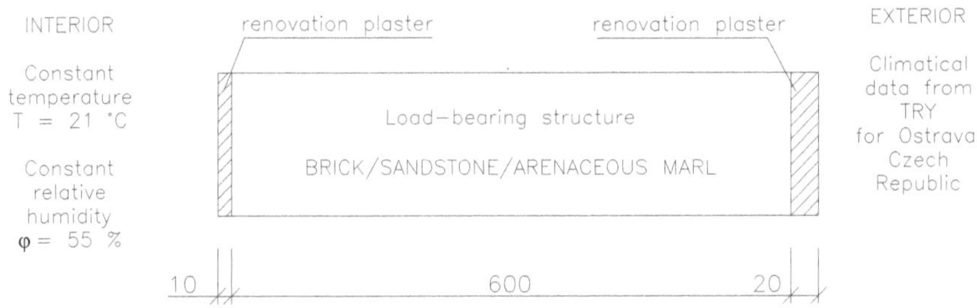

Figure 1: Scheme of studied envelope.

Three different kinds of materials of load-bearing structure were considered – ceramic brick, sandstone and arenaceous marl – together with two commercial renovation plasters (CRP_1 and CRP_2) and two laboratory developed systems of renovation plasters which are supposed to be produced commercially in the near future (LRP_1 and LRP_2).

The thickness of the load-bearing structure was 600 mm for all materials; the thickness of renovation plaster was 10 mm for interior side and 20 mm for exterior side. The investigated point was placed into the exterior plaster, 2 mm from the exterior surface in order to determine the number of freeze/thaw cycles.

3 Experimental

In the experimental work, the frost resistance of the chosen plasters was investigated under laboratory conditions according to ČSN 72 2452 "Testing of frost resistance of mortar". The specimen had dimensions of 40×40×160 mm. After removal from the moulds they were placed into open boxes and humidified once per day until the end of the 28-days curing period. Temperature in the laboratory was 21±1°C, relative humidity was 45±5%.

In order to measure freeze-thaw resistance of renovation renders, the water saturated testing specimens were cyclically frosted and defrosted until their damage got apparent. Three testing specimens of every plaster were put into water having temperature of 20±1°C until their full saturation. After that, the specimens were removed from water and then they were wiped by paper towel and put into plastic bag. Packed specimens were put into a freezing box for at least 4 hours at a temperature of –20±3°C. After removal, the specimens were kept in the laboratory with a temperature of 20±1°C for 2 hours, and then were put into water for 16 hours. This cycle was repeated until visible damage of specimens. The picture of plasters after 30 freeze-thaw cycles is shown in fig. 2. A little surface destruction of both coats is apparent, but the surface still seems to be compact. The overall freeze-thaw durability is shown in table 1. Commercial plasters (CRP1 and CRP2) after different number of freeze-thaw cycles are shown in fig. 2.

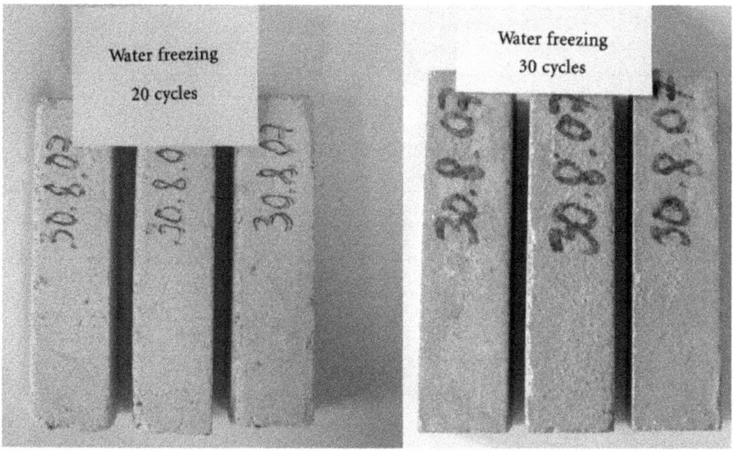

Figure 2: Commercial plasters after 20/30 freeze-thaw cycles (CRP_1 on the left, CRP_2 on the right).

Table 1: Number of freeze-thaw cycles causing apparent damage of renovation renders.

	CRP_1	CRP_2	LRP_1	LRP_2
Number of freeze-thaw cycles	30+	20	15	12

4 Computational

4.1 Materials characteristics

Basic parameters of materials involved in computer simulations are presented in tables 2 and 3, where ρ is the bulk density, ψ the porosity, μ the water vapour diffusion resistance factor, w_{hyg} the hygroscopic moisture content by volume, c the specific heat capacity, λ_{dry} the thermal conductivity in dry conditions, λ_{sat} the thermal conductivity in water saturated conditions.

Table 2: Material parameters of renovation renders.

	CRP_1	CRP_2	LRP_1	LRP_2
ρ [kg/m^3]	1296	1384	1687	1694
ψ [%]	50.2	47.3	34.1	34.4
c_{dry} [J/kgK]	1073	1028	877	877
c_{sat} [J/kgK]	1545	1719	1311	1311
$\mu_{dry\ cup}$ [-]	23.32	12.60	29.03	31.48
$\mu_{wet\ cup}$ [-]	6.67	7.10	11.57	14.12
λ_{dry} [W/mK]	0.293	0.366	0.886	0.836
λ_{sat} [W/mK]	1.383	1.186	2.190	2.267
w_{hyg} [m^3/m^3]	0.0932	0.0307	0.0493	0.0437

Table 3: Material parameters of load-bearing structure.

	Brick	Sandstone	Arenaceus marl
ρ [kg/m^3]	1670	1809	1400
ψ [%]	37.5	31.0	63.1
c [J/kgK]	1202	850	837
μ [-]	19.5	7.5	5.0
λ_{dry} [W/mK]	0.500	1.100	0.710
λ_{sat} [W/mK]	0.648	3.650	0.890
κ [m^2/s]	$2.65 \cdot 10^{-7}$	$2.5 \cdot 10^{-6}$	$1.0 \cdot 10^{-8}$
w_{hyg} [m^3/m^3]	0.0480	0.00567	0.01

All parameters were measured in the Laboratory of Transport Processes of the Department of Materials Engineering and Chemistry, Faculty of Civil Engineering, Czech Technical University in Prague [12]. The adsorption and desorption isotherms of studied renovation plasters are shown in fig. 3.

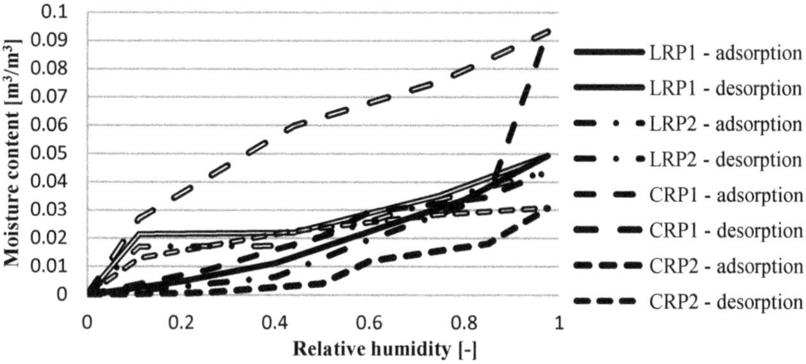

Figure 3: Adsorption and desorption isotherms of studied renovation plasters.

In fig. 4, the moisture diffusivity of the studied plasters is presented; this was obtained for the absorption phase only. Contrary to the adsorption isotherms, where desorption curves can be obtained by common experimental techniques, the desorption curve of moisture diffusivity had to be set empirically as for its experimental determination no quite reliable techniques are available at present. Based on the results of experiments and computational analyses described in [6], it was estimated to be one order of magnitude lower than the absorption curve.

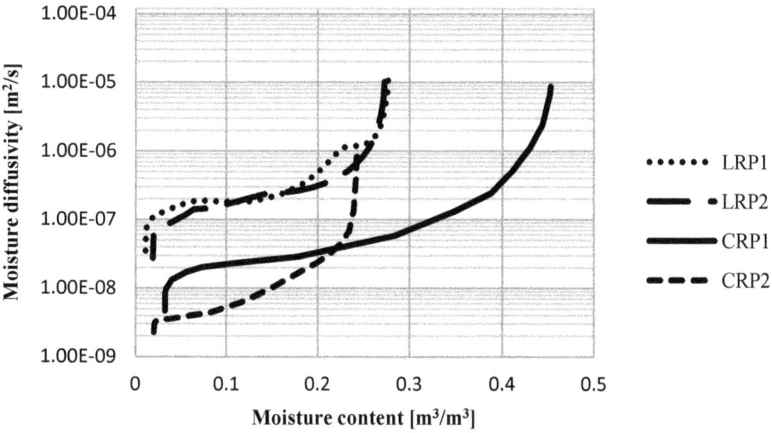

Figure 4: Moisture diffusivity of studied renovation plasters.

4.2 Computer simulation tool and mathematical model

The computational analysis was performed using computer code HEMOT [7], which was developed at the Department of Materials Engineering and Chemistry, Faculty of Civil Engineering, Czech Technical University in Prague in order to support the investigation of the coupled heat, moisture and salt transport in porous building materials. It enables simulation of thermal and hygric behaviour of structural building details in 1-D or 2-D arrangement. The building of the code is based on the application of the general finite element computer simulation tool SIFEL (Simple Finite Elements) developed at the Department of Mechanics, Faculty of Civil Engineering, Czech Technical University in Prague [8]. The Kunzel's diffusion model was used for the description of coupled heat and moisture transport [9]:

$$\frac{d\rho_v}{d\varphi}\frac{\partial\varphi}{\partial t} = div\left[D_\varphi grad\varphi + \delta_p grad(\varphi p_s)\right] \tag{1}$$

$$\frac{dH}{dT}\frac{\partial T}{\partial t} = div(\lambda gradT) + L_v div\left[\delta_p grad(\varphi p_s)\right] \tag{2}$$

where ρ_v is the partial density of moisture, φ the relative humidity, δ_p the permeability of water vapour, p_s the partial pressure of saturated water vapour, H the enthalpy density, L_v the heat of evaporation, λ the thermal conductivity, T the temperature,

$$D_\varphi = D_w \frac{d\rho_v}{d\varphi} \tag{3}$$

is the liquid moisture diffusivity coefficient, D_w the capillary transport coefficient.

Since the computational code involves the effect of hysteresis of moisture transport and storage parameters, which brings the computations nearer to reality, it was necessary to describe the path between adsorption and desorption phase. For that purpose an empirical procedure was chosen which follows Pedersen's hysteretic model [10],

$$\xi = \frac{a_d\left(w_p - w_a\right)^2 \xi_d + a_a\left(w_p - w_d\right)^2 \xi_a}{\left(w_d - w_a\right)^2}, \tag{4}$$

where w_p is the value of moisture content from the previous calculation step, w_a and w_d are the values of moisture content for adsorption and desorption cycles, respectively, ξ_a and ξ_d the corresponding values of tangent adsorption and desorption at points w_a and w_d, a_a and a_d the corresponding correction coefficients.

For describing the path between adsorption and desorption phase of moisture diffusivity, a modification was needed in order to express the hysteretic effect in more accurate way (see [11] for more details).

As boundary conditions in the exterior, climatic data for Ostrava were used, Czech Republic, obtained using Meteonorm software, version 6.1, which is a meteorological database and computer program for climatological calculations for every location on the globe.

5 Results of computer simulation

The results of computational simulations performed by the HEMOT code with the input parameters and boundary conditions presented in the previous sections are presented in figs 5–10. All the performed computations were for a period of five years. The results are given for the last year of the simulation.

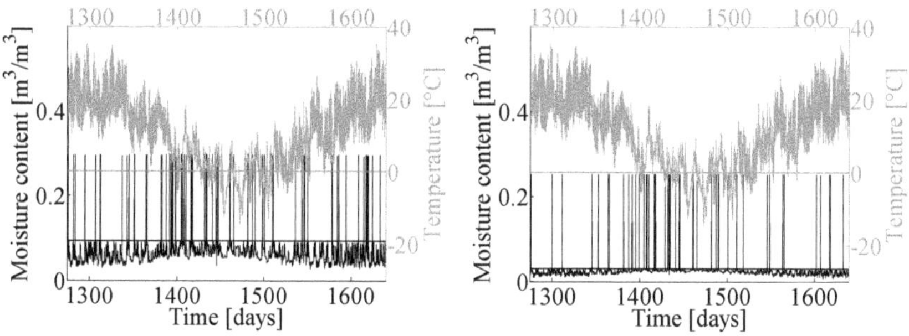

Figure 5: Time behaviour of temperature and moisture content in renovation plasters on load-bearing structure made of ceramic brick; CRP_1 on the left, CRP_2 on the right.

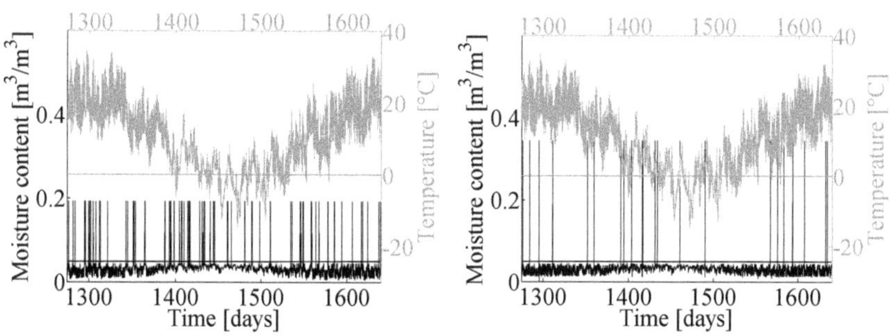

Figure 6: Time behaviour of temperature and moisture content in renovation plasters on load-bearing structure made of ceramic brick; LRP_1 on the left, LRP_2 on the right.

5.1 Load-bearing structure made of ceramic brick

The differences between commercial plasters (CRP_1 and CRP_2) and laboratory-prepared plasters (LRP_1 and LRP_2) are quite significant. While commercial renovation plasters CRP_1 and CRP_2 exhibit 4 and 5 freeze-thaw cycles, respectively, both laboratory-prepared renovation plasters showed only 1 freeze-thaw cycle.

5.2 Load-bearing structure made of sandstone

A similar situation as in brick masonry occurred in the sandstone masonry. Both laboratory-prepared plasters showed better results than commercial plasters. There occurred only 1 freeze-thaw cycle in LCP_1, 2 freeze-thaw cycles in LCP_2, 3 freeze-thaw cycles were discovered in CRP_1 and 4 cycles in CRP_2.

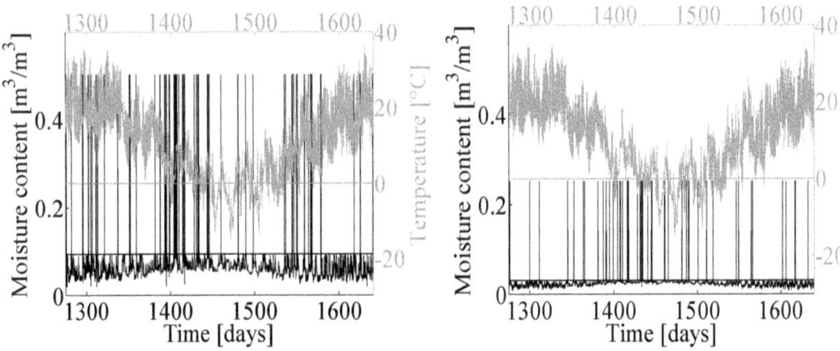

Figure 7: Time behaviour of temperature and moisture content in renovation plasters on load-bearing structure made of sandstone; CRP_1 on the left, CRP_2 on the right.

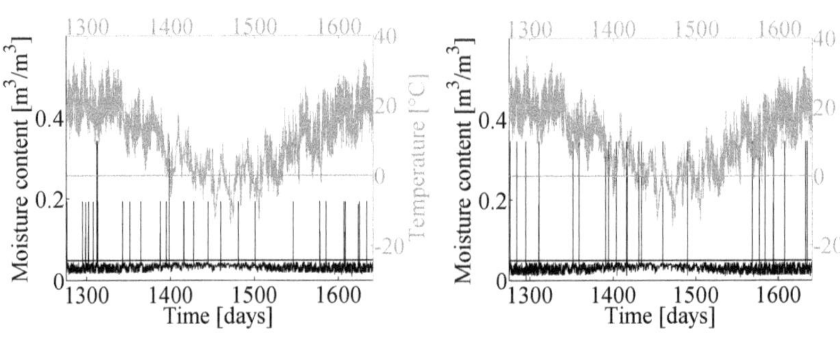

Figure 8: Time behaviour of temperature and moisture content in renovation plasters on load-bearing structure made of sandstone; LRP_1 on the left, LRP_2 on the right.

5.3 Load-bearing structure made of arenaceous marl

The worst results for all tested renovation renders were calculated for masonry made of arenaceous marl. Both laboratory-prepared renovation plasters showed 4 freeze-thaw cycles, the commercial plasters exhibited one and two cycles more, respectively (5 cycles for CRP_1 and 6 cycles for CRP_2).

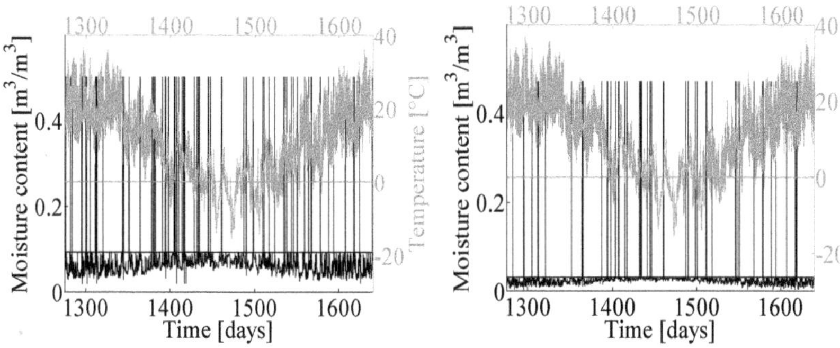

Figure 9: Time behaviour of temperature and moisture content in renovation plasters on load-bearing structure made of arenaceous marl; CRP_1 on the left, CRP_2 on the right.

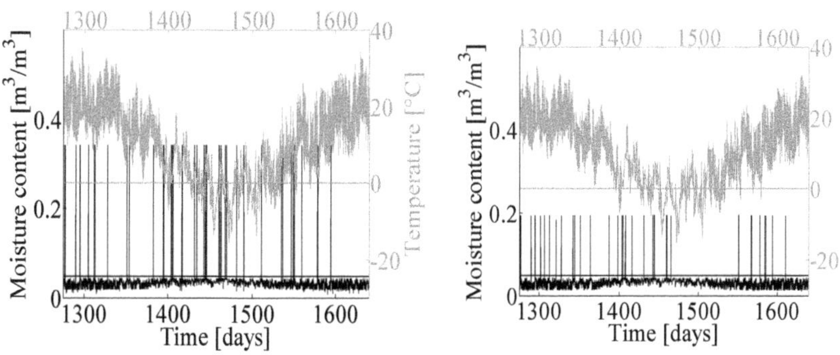

Figure 10: Time behaviour of temperature and moisture content in renovation plasters on load-bearing structure made of arenaceous marl; CRP_1 on the left, CRP_2 on the right.

6 Discussion

In this paper, phase changes of water were assumed as the only factor affecting the service life of innovative plasters. In reality, there are more factors playing the key role in the damage of surface layers of renovation renders, such as salt transport, microbiological damage and acid rains. On the other hand, this simplification based on separation of the particular effect of damage (frost resistance) from a variety of other effects can provide valuable information indeed.

All performed computations proved that freeze-thaw cycles may occur in surface layers of renovation renders, even when the reference year is taken into account. In table 4, all results calculated using computer simulations are summarized.

Table 4: Summary of freeze/thaw cycles appearance during a reference year.

	LRP_1	LRP_2	CRP_1	CRP_2
Ceramic brick	1	1	4	5
Sandstone	1	2	3	4
Arenaceous marl	4	4	5	6

The best results were achieved for LRP_1 but all investigated plasters fared well. The most significant differences between studied plasters were discovered in brick masonry. The other types of masonry resulted in very similar appearance of freeze-thaw cycles.

Taking into account the frost resistance of each renovation render (table 1), its service life can be predicted. From this point of view, it may be recommended to use LRP_1 or LRP_2 for brick masonry, LRP_1 or CRP_1 for sandstone masonry and CRP_1 for masonry made of arenaceous marl. An interesting finding is that both commercially available products CRP_1 and CRP_2 sold for a similar price showed very different results. The average service life of CRP_1 was approximately two times higher than that of CRP_2, with respect to frost resistance.

7 Conclusions

The computational results presented in this paper dealing with the hygrothermal performance of innovative renovation renders showed that a proper choice of materials for renovation, with respect to the material of the load-bearing structure, could significantly extend the service life of a renovated building envelope. In general, the best results were achieved for brick masonry, whereas the worst results were obtained for masonry made of arenaceous marl. It is important to say that the durability of renovation renders is not only affected by cyclic freezing and wetting. A crucial contribution to reducing the service life of all types of renovation renders comes from salt accumulation and crystallization due to their high open porosity. Therefore, in future work, the mathematical model should be extended to include salt transport as well.

Acknowledgement

This research has been supported by the Czech Science Foundation, under grant No 103/09/0780.

References

[1] Kočí, J., Kočí, V., Maděra, J., Rovnaníková, P. & Černý, R., Computational analysis of hygrothermal performance of renovation renders. *Advanced Computational Methods and Experiments in Heat Transfer XI.* WIT Press: Southampton, pp. 267–277, 2010.

[2] Maděra, J., Kočí, V., Vejmelková, E., Černý, R., Rovnaníková, P., Ondráček, M. & Sedlmajer, M., Influence of material characteristics of concrete and thermal insulation on the service life of exterior renders. *Computational Methods and Experimental Measurements*, WIT Press: Southampton, pp. 13–23, 2009.
[3] Lu, G., Lu, G.Q. and Xiao, Z.M., Mechanical properties of porous materials. *Journal of Porous Materials*, **6**, pp. 359 –368, 1999.
[4] Larbi, J.A., Microscopy applied to the diagnosis of the deterioration of brick masonry. *Construction and Building Materials*, **18**, pp. 299–307, 2004.
[5] Mendes, N. & Philippi, P.C., A method for predicting heat and moisture transfer through multilayered walls based on temperature and moisture content gradients. *International Journal of Heat and Mass Transfer,* **48**, pp. 37–51, 2005.
[6] Pel L., Černý R. & Pavlík Z., Moisture and Ion Transport. WP5 2-Years Report of the EU 6th Program Project SSPI-CT-2003-501571. TU Eindhoven, Eindhoven, 2006.
[7] Černý R., Complex System of Methods for Directed Design and Assessment of Functional Properties of Building Materials: Assessment and Synthesis of Analytical Data and Construction of the System. CTU: Prague, pp. 192–201, 2010.
[8] Kruis, J., Koudelka T. & Krejčí, T., Efficient computer implementation of coupled hydro-thermo-mechanical analysis. *Mathematics and Computers in Simulation*, **80(8)**, pp. 1578–1588, 2010.
[9] Künzel, H. M., *Simultaneous Heat and Moisture Transport in Building Components*, PhD Thesis, IRB Verlag Stuttgart, 1995.
[10] Pedersen, C. R., *Combined Heat and Moisture Transfer in Building Constructions*, PhD Thesis, Report 214. Thermal Insulation Laboratory, TU Denmark, 1990.
[11] Černý, R., Maděra, J., Kočí, J. & Vejmelková, E., Heat and moisture transport in porous materials involving cyclic wetting and drying. *Computational Methods and Experimental Measurements XIV*, WIT Press: Southampton, pp. 3–12, 2009.
[12] Vejmelková, E., Keppert, M., Máca, P. & Černý, R., Mechanical, hygric and thermal properties of innovative renovation renders. *Structural Repairs and Maintenance of Heritage Architecture XII*, WIT Press: Southampton, pp. 555–563, 2011.

Feasibility of integral water repellent admixtures in low pressure compatible injected fill grouts

C. Citto, A. E. Geister & D. W. Harvey
Atkinson-Noland & Associates, Inc., USA

Abstract

For almost two decades, compatible injected fill (CIF) grouts have been used in the restoration and repair of historic masonry structures. CIF materials are installed using a low-pressure injection process and often serve the dual purposes of structural strengthening and enhancement of moisture resistance. In the past, improved weather resistance was primarily the result of the CIF material filling voids in collar, head, and bed joints through which moisture could penetrate. However, the CIF materials were not specifically formulated for moisture resistance. The weather resistance of CIF installations could be significantly improved by increasing the water repellency of the CIF material. However, if this innovation comes at the expense of significant changes to other material properties, it may not be a viable approach. This paper presents the ability of injected CIF material to improve the moisture resistance of masonry walls through a series of water penetration test results, and also describes a comparative study of CIF material properties in formulations with and without integral water repellent (IWR) admixture. Material properties relevant to typical CIF installations were evaluated using standard laboratory test procedures. Samples were tested for plastic properties, compressive strength, flexural bond strength, shear bond strength, vapor permeability and water repellency. Tested samples used the same proprietary blend of cementitious materials, aggregate and admixtures. The effects of IWR on material properties are discussed as well as overall feasibility of IWR admixtures in CIF materials.

Keywords: building envelope, CIF, grout injection, integral water repellent, masonry walls, moisture resistance, water penetration, water testing.

1 Introduction

Compatible injected fill (CIF) materials are a valuable tool used in the restoration and repair of historic masonry structures. During the development of the CIF approach, an extensive experimental program was conducted by Atkinson and Schuller [1] to evaluate injection grouting procedures, different mix designs and the effect on structural behavior of CIF injections. Nevertheless, experimental work by Albert [2] showed that water penetration through masonry cavity walls could be significantly reduced when filling the cavity by injection grouting. However, CIF materials were not specifically formulated for moisture resistance. The weather resistance of CIF installations could be further improved by increasing the water repellency of the CIF material. This paper illustrates results of an experimental investigation carried out to evaluate the influence of integral water repellant (IWR) admixture to relevant material properties of CIF. The testing performed on the CIF specimens was intended only as a qualitative comparison between mixes with and without IWR.

2 Water penetration testing of masonry walls

Two standardized test methods are commonly used to evaluate and compare masonry water penetration performance: the laboratory test in accordance with ASTM E514, *Standard Test Method for Water Penetration and Leakage Through Masonry* [3], and the corresponding field test in accordance with ASTM C1601, *Test Method for Field Determination of Water Penetration of Masonry Wall Surfaces* [4]. Both methods use a chamber mounted on a masonry surface to simulate a severe rainstorm producing approximately 5½ inches (140 mm) of rain per hour and 62.5 mph (27.9 m/s) wind for 4 hours. These methods are commonly used to evaluate and compare masonry water penetration performance.

ASTM E514 results by Albert [2] showed that injecting the cavity of brick veneer laboratory test panels with a conventional, non-IWR CIF reduced water penetration in a 2-wythe brick wall from 12.4 L/hr to 0.08 L/hr, a brick veneer on concrete block wall from 5.83 L/hr to 0.64 L/hr, and brick veneer on steel stud drywall from 3.63 L/hr to zero.

ASTM C1601 testing was performed by the authors on a single-wythe, split face concrete masonry building before and after the hollow cells in a section of wall were filled by injection grouting with a conventional, non-IWR CIF. Injection grouting reduced the water penetration rate by 68%, from 60 L/hr to 19 L/hr. Testing in numerous multi-wythe brick masonry walls has generally shown dramatic reductions in moisture infiltration after installation of CIF without IWR.

3 Comparative experimental program

The main purpose of this study was to investigate whether the water repellency of CIF could be improved with the addition of IWR without negatively affecting other CIF properties in a significant way. The addition of IWR should not compromise the capacity of the CIF material to fill small voids and cracks without segregation and bond to the surrounding masonry, should be compatible with the masonry being injected and have adequate strength. To this end, three sets of specimens were prepared and tested, the first being designated the "base" mix without IWR, the second being a mix containing a liquid oleic acid IWR, and the third containing a powder calcium stearate IWR. Dosage of IWR was based on manufacturer's recommendations for mortar, which relate IWR to total binder content by weight. All other components and aspects of CIF mix formulation were kept identical. Testing involved the evaluation of plastic properties, hardened mechanical properties, and water resistivity of the three mixes.

3.1 Plastic properties

The first step of the experimental program was to characterize the three different mixes through the evaluation of their plastic properties. Fluidity was measured in accordance with ASTM C939, *Standard Test Method for Flow of Grout for Preplaced-Aggregate Concrete (Flow Cone Method)* [5]. After 3 minutes of electric mixing, the flow rate of each mix was tested using a 0.5 inch (12.7 mm) diameter flow cone, and the ratio of water to dry mix was adjusted accordingly to obtain consistency between the mixes before observing other properties and casting specimens. CIF mix stability was evaluated for each mix in accordance with ASTM C940, *Standard Test Method for Expansion and Bleeding of Freshly Mixed Grouts for Preplaced-Aggregate Concrete in the Laboratory* [6]. This test requires the evaluation of volume changes and accumulation of bleed water in a graduated cylinder. To compare the stability of the mix formulations, a more severe variation of ACI 423.9M-10, *Test Method for Bleed Stability of Cementitious Post-Tensioning Tendon Grout* [7] was used to measure water separation after 10 minutes under 10 psi (68.9 kPa) pressure in a Gelman pressure cell. Stability is an important property of the CIF as segregation of the constituents during the injection process can result in incomplete filling of voids or inconsistent bonding and strength properties.

3.2 Mechanical properties

Different tests were conducted to evaluate the influence of the IWR on CIF mechanical properties. Comparison of compressive strength was determined by casting and testing CIF specimens in accordance with ASTM C1019, *Standard Test Method for Sampling and Testing Grout* [8]. In this test method, molds are made from masonry units to subject the CIF specimens to absorption conditions similar to those experienced by CIF material in the wall. Three samples were obtained from each mix. Flexural bond strength was determined in accordance with ASTM C1072, *Standard Test Method for Measurement of Masonry Flexural Bond Strength* [9]. The stack bond test prisms were created by filling the bed joints between bricks with CIF. A set of porous historic brick and a set of dense modern solid brick were used for each mix type to subject the CIF to different absorption conditions in the flexural bond strength test. Comparison of shear bond strength was determined by using a modified variation of California Test 644, *Core Test for Shear Bond* [10]. A two-wythe brick wall panel was constructed, and the cavity between wythes was filled with CIF. 28 days after casting CIF, cores were removed from the panel and the brick-CIF interface was tested in shear. Three cores were tested for each mix formula.

3.3 Water repellency

Water repellency properties were evaluated using adaptations of National Concrete Masonry Association (NCMA) tests for characteristics of concrete masonry units (CMU) with IWR. Water absorption was measured in accordance with variations of NCMA Methods CMU-WR1, *Standard Test Method for Water Droplet and Water Stream Tests of Concrete Masonry Units* [11] and CMU-WR3, *Standard Test Method for Assessing Water Uptake Potential of Concrete Masonry Units* [12] were used. Since these tests are intended to be performed on portions of CMU face shells, CIF specimens were cast with dimensions similar to a CMU face shell, as seen in fig. 1, for this purpose. Three samples were made using the base mix and three using each IWR mix, and all tests were performed after a minimum cure time of 28 days. In the water droplet test, multiple drops of water were placed on the hardened CIF surface and the ability of the material to resist the absorption of water was evaluated over time based on visual observation. In the water uptake test, the oven-dried specimen is placed in 0.1 inch (3 mm) of water and the water absorption is evaluated by measuring the change in weight of the specimen for the next 24 hours.

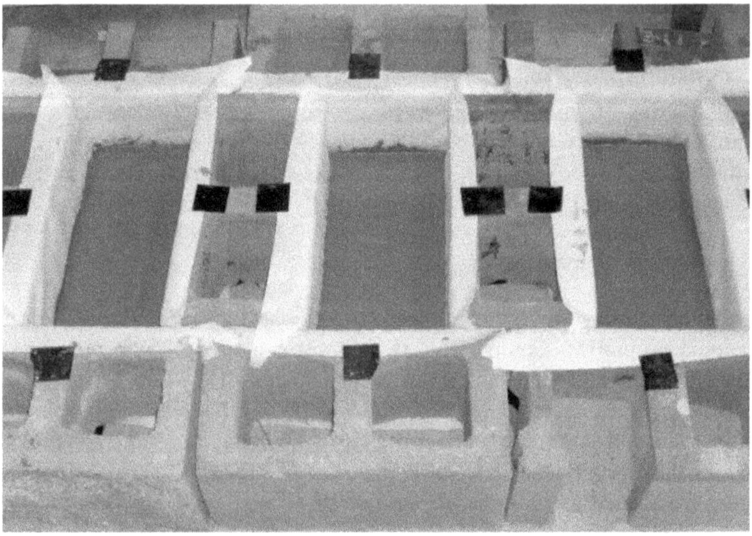

Figure 1: CIF specimens cast for water repellency testing.

3.4 Water vapor permeability

Water vapor transmission (WVT) testing was conducted for each mix in accordance with ASTM E96, *Standard Test Methods for Water Vapor Transmission of Materials* [13]. Specimens for the WVT testing were prepared by sectioning CIF samples cast in absorptive brick molds with a water-cooled masonry saw. A minimum of four samples were obtained from each mix with dimensions of approximately 2.4 inches (60 mm) long by 2.4 inches (60 mm) wide by 0.6 inches (15 mm) thick. Samples were inserted into square plastic containers and sealed with silicone at all perimeter edges. Water was introduced into the wet cup by hypodermic needle and the puncture then sealed with silicone. All samples were placed in an environmental chamber with conditions maintained at 90°F (32°C) and relative humidity of 50%. The temperature was controlled by a thermostat and heating element and the relative humidity was maintained by an open container of saturated Magnesium Nitrate Hexahydrate placed in the environmental chamber. This salt provides an equilibrium relative humidity of 50% at 90°F (32°C).

4 Experimental results

4.1 Plastic properties

Properties measured after mixing included flow rate, fluid density, expansion in a graduated cylinder and water separation under an applied pressure of 10 psi (68.9 kPa) for 10 minutes. The results of these plastic properties are presented in table 1. Note that slightly more water was required in both IWR mixes to achieve a similar flow rate as the Base mix. A very small amount of water separation was observed within 10 minutes under the applied pressure in all three mixes. CIF density decreased slightly with the addition of IWR; the amount of initial expansion observed in the base mix was reduced to half in the calcium stearate IWR mix and to zero in the oleic acid IWR mix.

Table 1: CIF mix plastic properties.

Mix designation	Flow cone time (sec)	Baroid density (g/cm^3)	Final expansion (%)	Final bleeding (%)	Water separation under pressure (ml)
Base mix	24	2.02	1.3%	0	1
Oleic acid IWR mix	27	2.00	0	0	1
Calcium stearate IWR mix	25	1.98	0.6%	0	1

4.2 Mechanical properties

CIF compressive strength was tested at an age of 28 days. Results are summarized in table 2. Compressive strength results of the IWR mix formulas were about 15% to 30% lower on average, possibly due to the additional water required to achieve the same flow rate as the base mix. The aim of CIF, however, is generally not to achieve high values of compressive strength, but rather to be compatible with the surrounding materials, which may have very low compressive strengths.

Table 2: CIF 28 day compressive strength test results.

	Base mix	Oleic acid IWR mix	Calcium stearate IWR mix
Compressive strength, psi (MPa)	9000 (62)	6900 (48)	7800 (54)
Standard deviation, psi (MPa)	110 (0.8)	210 (1.4)	280 (1.9)

Flexural bond strength was tested at an age of 28 days using the bond wrench apparatus described in ASTM C1072 [9]. Results, summarized in table 3, saw an overall decrease in flexural bond strength of dense modern brick with the use of IWR, while still achieving very high values. More variation was observed in the historic brick results, because failure generally occurred within the brick, indicating the CIF bond was stronger than the flexural tensile strength of the brick material.

Table 3: Flexural bond strength results.

	Base mix	Oleic acid IWR mix	Calcium stearate IWR mix
Flexural bond strength with modern brick, psi (MPa)	340 (2.4)	250 (1.7)	190 (1.3)
Standard deviation, psi (MPa)	210 (1.5)	80 (0.5)	10 (0.1)
Flexural bond strength with historic brick, psi (MPa)	54 (0.37)	22 (0.15)	83 (0.57)
Standard deviation, psi (MPa)	44 (0.31)	21 (0.14)	50 (0.34)

Shear bond strength was tested at an age of 28 days using a modification of California Test 644 [10]. Results are summarized in table 4. Test results were higher with the addition of IWR, indicating that its use is not likely to have a negative effect on shear bond strength.

Table 4: Shear bond strength test results.

	Base mix	Oleic acid IWR mix	Calcium stearate IWR mix
Shear strength, psi (MPa)	110 (0.75)	115 (0.8)	150 (1.0)
Standard deviation, psi (MPa)	36 (0.25)	16 (0.11)	35 (0.24)

4.3 Water repellency

The addition of IWR appears to improve the water repellency performance of the CIF in the water droplet test. Although this test is subjective in nature, the results showed significantly better performance by the IWR mix than the base mix. Using the nomenclature of the observation scale given in the test method, after 30 minutes, over 70% of the base mix droplets were considered "Totally absorbed," and after 60 minutes, over 85% were considered "Dry." All of the oleic acid IWR mix droplets were either "Standing" or "Partially absorbed" after 60 minutes. 60% of the calcium stearate mix droplets were considered "Partially absorbed" after 60 minutes, one third were considered "Totally absorbed" and the remaining droplets were considered "Dry."

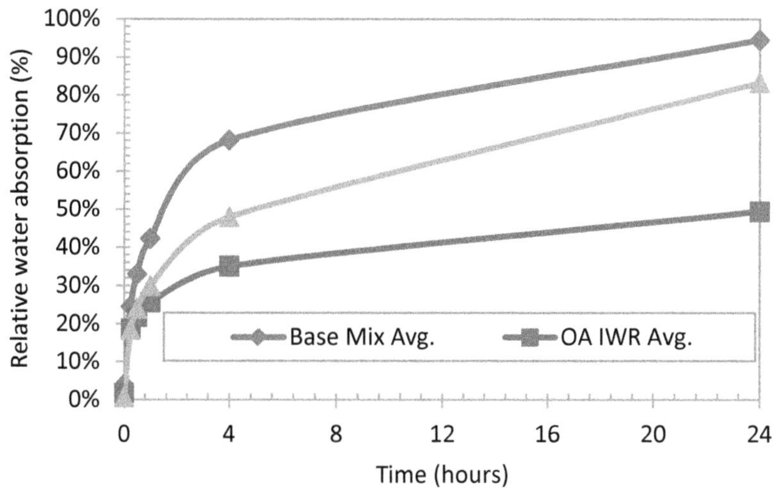

Figure 2: Relative water uptake results.

Water uptake testing was performed on CIF samples in accordance with NCMA WR3 [12]. Average results in fig. 2 show an improvement in water uptake resistance with the use of IWR. Water uptake performance was improved with the addition of IWR. While 4 out of 5 base mix specimens reached 90% of total water absorption before the 24 hour measurement, only 2 out of 5 of the calcium stearate IWR specimens reached 90% by the 24 hour measurement, and none of the IWR mix specimens reached 50% by the 24 hour measurement.

4.4 Water vapor permeability

Results of water vapor transmission test are shown in table 5. Similar behavior in terms of change in weight over time was observed for all three mixes. However, the oleic acid IWR mix experienced a reduction in WVT rate, while the calcium stearate IWR mix experienced an increase in WVT rate. These relatively small changes in water vapor permeability are unlikely to result in substantially different wall performance in most applications.

Table 5: Water vapor permeability test results.

	Base mix	Oleic acid IWR mix	Calcium stearate IWR mix
Permeability, Perm·inch (g/Pa·s·m)	4.2 (6.1×10^{-9})	3.3 (4.8×10^{-9})	6.1 (8.8×10^{-9})
Standard deviation, Perm·inch (g/Pa·s·m)	0.8 (1.1×10^{-9})	0.6 (0.8×10^{-9})	0.6 (0.8×10^{-9})

5 Conclusions

A review of the use of compatible injected fill (CIF) grout for the use of increasing the water resistance of masonry walls was presented. A series of tests performed before and after filling the wall cavity with CIF showed that water penetration could be significantly reduced even though CIF had not been specifically formulated for water resistance.

The influence of integral water repellent on material properties of CIF was evaluated. Tests were carried out on CIF specimens cast with and without integral water repellent. Samples were tested for plastic properties, compressive strength, flexural bond strength, shear bond strength, vapor permeability, and water repellency. The comparative study showed that plastic properties are slightly affected by IWR as a small amount of additional water was needed to achieve the required flow rate. Furthermore, the water repellency performance of the CIF was improved with the addition of IWR. Lower values of compressive strength were observed in each IWR mix, with an average reduction in the range of 15% to 30%, however, compatibility with surrounding materials is generally more important in CIF installations than high values of compressive strength. Flexural bond strength results were variable, but still achieved very high bond strength values on dense modern brick, and lower strength results on softer historic brick resulting from failure of the brick material itself, rather than the CIF bond. Shear bond strength appeared to be unaffected by the use of IWR. The water vapor transmission rate of the oleic acid IWR mix decreased, while the mix containing calcium stearate IWR experienced an increase in water vapor transmission rate. In general, it appears that the use of IWR in CIF installations would likely be an effective approach to increasing water repellency of masonry walls.

References

[1] Atkinson, R.H. & Schuller, M.P., Evaluation of injectable cementitious grouts for repair and retrofit of masonry. *Masonry: Design and Construction, Problems and Repair, ASTM STP 1180,* eds. J. M. Melander & L. R. Lauersdorf, American Society for Testing and Materials, Philadelphia, 1993.

[2] Albert, S.D., *The Effectiveness of Grout Injection in Preventing Water Penetration Through Masonry Cavity Walls,* Supported by National Science Foundation Grant No. MSS-9114511, 1996.

[3] ASTM E514. *Standard Test Method for Water Penetration and Leakage Through Masonry*, Annual Book of ASTM Standards, Vol. 04.05. ASTM, West Conshohocken, PA, 2009.

[4] ASTM C1601. *Test Method for Field Determination of Water Penetration of Masonry Wall Surfaces*, Annual Book of ASTM Standards, Vol. 04.05. ASTM, West Conshohocken, PA, 2010.

[5] ASTM C939. *Standard Test Method for Flow of Grout for Preplaced-Aggregate Concrete (Flow Cone Method)*, Annual Book of ASTM Standards, Vol. 04.02. ASTM, West Conshohocken, PA, 2010.

[6] ASTM C940. *Standard Test Method for Expansion and Bleeding of Freshly Mixed Grouts for Preplaced-Aggregate Concrete in the Laboratory*, Annual Book of ASTM Standards, Vol. 04.02. ASTM, West Conshohocken, PA, 2003.

[7] ACI 423.9M-10. *Metric Test Method for Bleed Stability of Cementitious Post-Tensioning Tendon Grout*, Joint ACI-ASCE Committee 423. American Concrete Institute, Farmington Hills, MI, 2010.

[8] ASTM C1019. *Standard Test Method for Sampling and Testing Grout*, Annual Book of ASTM Standards, Vol. 04.05. ASTM, West Conshohocken, PA, 2009.

[9] ASTM C1072. *Standard Test Method for Measurement of Masonry Flexural Bond Strength*, Annual Book of ASTM Standards, Vol. 04.05. ASTM, West Conshohocken, PA, 2010.

[10] California Test 644, *Method of Test for Shear Strength of Brick Cores*, California Department of Transportation, 2000.

[11] NCMA Method CMU-WR1-09. *Standard Test Methods for Water Droplet and Water Stream Tests of Concrete Masonry Units,* National Concrete Masonry Association, 2009.
[12] NCMA Method CMU-WR3-09. *Standard Test Method for Assessing Water Uptake Potential of Concrete Masonry Units,* National Concrete Masonry Association, 2009.
[13] ASTM E96. *Standard Test Methods for Water Vapor Transmission of Materials*, Annual Book of ASTM Standards, Vol. 04.06. ASTM, West Conshohocken, PA, 2010.

Assessment of ancient masonry slender towers under seismic loading: dynamic characterization of the Cuatrovitas tower

P. Pineda & A. Sáez
School of Architecture, University of Seville, Spain

Abstract

The Cuatrovitas tower is a XIIth century almohad minaret located in the province of Seville (Spain) and it is considered the best preserved almohad religious building in the Iberian Peninsula. As it is placed in a seismic area, it is crucial for its preservation to evaluate its dynamic response under earthquake loading and to assess the safety level in its present state of conservation. In this paper a number of three-dimensional linear and non-linear finite element models with different levels of complexity and simplifications are developed, using 3-D solid elements or 3-D beams elements. All the models assume that the masonry structure is homogeneous and the material non-linear behaviour – including crushing and cracking – is simulated by means of different constitutive models. Subsequent non-linear static and non-linear dynamic analyses are performed. Previous static and time-history dynamic analyses with a simplified elastic material model are evaluated to calibrate the non-linear response, and to take into account that crack opening may introduce numerical instabilities. Comparison among the different models is thoroughly discussed, in particular as predicted local and global collapse mechanisms are concerned, in order to evaluate the suitability, accuracy and limitations of each analysis. To conclude, a general methodology is proposed to assess safety and to improve seismic resistance of this and other similar cultural heritage buildings.
Keywords: ancient masonry tower, almohad architecture, non-linear dynamic analysis, earthquake loading.

1 Introduction

Historic masonry towers placed in seismically active regions could be severely damaged by earthquakes. Indeed, these structures are able to resist gravitational actions, but as they were not explicitly designed to withstand seismic effects, they show particularly weakness with regard to horizontal loadings induced by dynamic excitations. The high vulnerability of these constructions under horizontal actions is mostly due to the absence of adequate structural connections, which leads to overturning collapse [1]. Regarding Almohad minarets, they are medieval towers of a type that is prone to suffer damages under earthquakes, mainly due to their slenderness, low shear strength and low ductility. Moreover, their possible lack of effective connections among structural elements – core, vaults and external walls – needs to be evaluated. Due to these reasons, it is crucial for their preservation to assess seismic safety in order to evaluate their dynamic response and, if necessary, to improve their structural strength. Indeed, prevention and rehabilitation can be successfully achieved only if diagnosis of the building is carefully analysed [2].

It is well known that analysis and evaluation of seismic reliability of masonry cultural heritage buildings is a difficult task, owing to uncertainties that mainly affect structural behaviour and mechanical material properties. The former includes lack of information on model definition – geometry, constraints, materials, constructive details – and the latter is focused on non-linear masonry behaviour and low tensile strength. Furthermore, if a comprehensive study of structural behaviour is going to be performed, accuracy and suitability of the analytical or numerical method selected are essential issues.

A contribution to dynamic characterization, lateral capacity and seismic assessment of Almohad minarets is provided in a single case study, revealing advantages or disadvantages of different numerical analyses – static, modal, linear transient, non-linear transient and non-linear static. Simplified and detailed models based on the finite element method are performed, and brittle non-linear behaviour of masonry is considered at the macro-level.

This research aims at predicting local and global collapse mechanisms, thus developing an accurate and practical method of analysis of dynamic response in Almohad constructions.

2 Historical survey

The Cuatrovitas medieval tower is a XIIth century minaret located in Bollullos de la Mitación, in the province of Seville, Spain, and nowadays it is part of the Nuestra Señora de Cuatrovitas Church (fig. 1). Furthermore, it is considered the best preserved Almohad religious construction erected in the Iberian Peninsula and, due to its proportions, sobriety and decoration it is a masterpiece of its architectonical style.

Minarets were introduced at the end of the second century of Islam in Abbasid Mesopotamia. They were attached to a mosque in order to signify the acceptation degree of the religious edifice as an institution in Islamic society.

In Almohad times, mosque towers had symbolic function too, since they were conceived as architectural statements of Islam during the Christian reconquest of Spain. The first of the great minarets of this period, the Kutubiyya tower, was erected in 1158 in Marrakesh, Morocco, at the left end of the quibla wall of

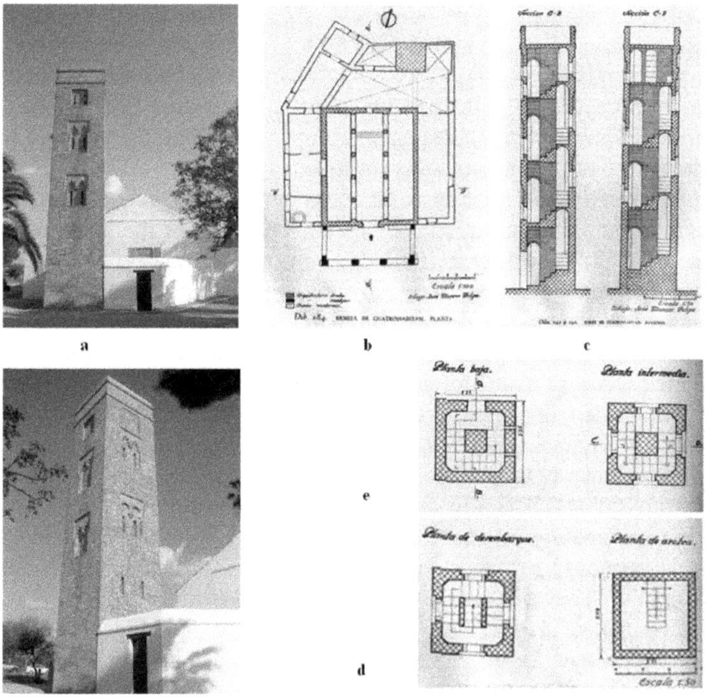

Figure 1: (a) General frontal view, and north and west façades; (b) Church plan section; (c) tower longitudinal section; (d) tower plan section. (Source: Catálogo arqueológico de la provincia de Sevilla.)

the mosque. When Seville became the Iberian capital of Almohads, the tower of its great mosque was constructed by Ahmad ibn Basu in 1184 on orders from Abu Ya'qub Yusuf. Unfortunately, the last of the more significant minarets belonging to this epoch, the tower at the mosque of Hassan at Rabat, is unfinished. In this historical context the analysed minaret was built. Although not mentioned in any sources, the minaret has been dated in 1175–1180 by stylistic studies [3]. Comparison with the aforementioned towers suggests that the Cuatrovitas Minaret has similar shape and style of decoration. Besides, all of them are located in active seismic areas. These monuments are essential in evolution of Islamic religious architecture as they were models for minarets built in the metropolitan centres of the Maghreb until Ottoman times [4].

As a reference of the value of the monument studied in this work, it is important to mention that it has been inscribed in the Spanish Heritage Monument Listing since 1931.

3 Geometrical and structural survey

The Cuatrovitas minaret is a free-standing tower situated next to the north façade of the church, walling in the *shan* – the original mosque court. This relevant architectural monument is of moderate size, 3.25×3.25 m^2 in lower plan, 3.10×3.10 m^2 in upper plan and it rises 14.8 m above the current ground level. Its plan-height relation is 1/4.21, complying with Almohad slenderness canons. The façade is decorated with rectangular panels and dual blind arches – five-foiled or horseshoe arches – framing small openings along the external walls, as it is in the nature of the purest Unitarian style.

Morphologically, the minaret is divisible into three structural parts: external walls, central core and barrel vaults (fig. 1). The average thickness of walls is 0.425 m, core cross-section is 0.90×0.90 m^2 and barrel vaults cross section is 0.14 m. The inner chamber consists of an anticlockwise staircase covered by horizontal barrel vaults, which ascend around the square central solid core. The whole structure is built of clay bricks and lime mortar joints, originally covered with painted plaster. The aforementioned building materials are essential in Islamic Andalusian constructions. Brick average size is 0.28×0.14×0.045 m^3 and average lime thickness is 0.035 m. A careful visual survey of the structure revealed irregular mortar-brick disposition in several zones, where bricks were substituted by mortar (fig. 2). Moreover the tower only exhibits superficial cracks. No information on the character of the foundation is available, but the structure is supposed to be embedded in the soil. To conclude it is important to mention that no severe damage is observed.

For a complete report on this building refer to [3]. In addition, a comprehensive survey of the archaeological features can be found in [5].

Figure 2: Aspects of the irregular mortar-brick disposition.

4 Static analysis

4.1 Linear material

The first analysis carried out focused on static assessment of the detailed three-dimensional finite element model, which allows the accurate consideration of all the essential structural features. The ANSYS finite element (FE) software was used to construct the model, and three-dimensional eight node solid elements, SOLID 45, were employed for masonry material [6]. The mesh consists of 65,152 elements, 52,188 nodes and 154,956 active degrees of freedom; average element size is 0.11m. The wall thickness was discretized with four elements. This model comprises the most significant structural parts, namely real load-bearing wall thickness – replicating the rectangular panels framing openings, which reduce the average cross section – openings, vaults and central core (fig. 3). The staircase self-weight was applied on the vaults, due to its constructive features. No information regarding physical material properties is available, thus the properties of contemporary structures in the region with the Cuatrovitas minaret were considered [7]. Furthermore, a smeared model with homogenized properties was developed and its linear elastic material properties were assumed as Young modulus $E = 1$ GPa, Poisson ratio $v = 0.2$, and specific weight $w = 17{,}000\ \mathrm{N/m^3}$.

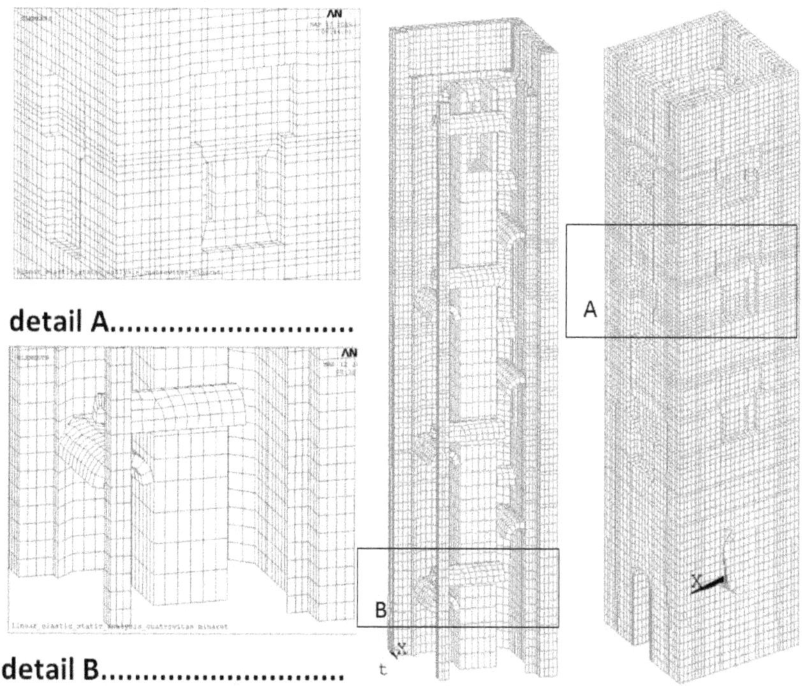

Figure 3: Finite element model mesh.

With regards to boundary conditions, the base of the tower was considered as completely constrained and the contribution of the adjoining church was assumed to be negligible.

This preliminary study provides valuable information both on global behaviour and on interaction among the structural parts. Indeed, the analysis of the structure under gravity loading yields significant data, such as stress distribution, weak elements of potential failure and displacements (table 1).

From the static numerical approach it may be observed that maximum compression level – 0.45 MPa – was reached by the basement, as expected, whereas maximum tensile stresses – 0.17 MPa – appeared in the vaults. Furthermore, high stress concentrations were obtained in the connections among structural parts. The compressive stresses are admissible, but as stresses higher than tensile strength were obtained, the use of non-linear material is advisable. Furthermore, those results allow one to conclude that connections between vaults and walls are confirmed to be a vulnerable part of the building. With regard to displacements, the maximum horizontal values were reached by the vaults and the core.

Table 1: Static analysis results.

Material	u_x (m)	u_y (m)	u_z (m)	σ_1 (MPa)	σ_3 (MPa)
Linear	0.67e–3	0.23e–3	0.0025	0.17	–0.45
Non-linear	0.68e–3	0.23e–3	0.0026	0.17	–0.36

4.2 Non-linear material

A second static analysis was performed on the same described three-dimensional finite element model, taking into account the material non-linear behaviour. The Drucker-Prager perfectly plastic criterion [8] and the Willam-Warncke failure surface [9] were employed in the model. It is important to mention that those criteria provide neither stiffness degradation of brittle material caused by successive plastic deformation, nor cracks resulting from low cycle fatigue [10, 11]. However, both theories yield accurate results with three-dimensional solid models, in particular when predicted cracking progression is concerned. Furthermore, the previous static analysis and the following non-linear analysis show that either crushing or plastic deformation due to high compressive levels was not found in the analysed structure. Prior literature on masonry structural analysis [10, 12–14] has used those criteria in order to determine the frontier between linear and non-linear behaviour.

Three-dimensional eight-noded solid isoparametric element, SOLID 65, was employed to model the brittle material. The element is capable of cracking in tension – in three orthogonal directions – and crushing in compression. The masonry was assumed to be initially isotropic, until either one of the tensile or the compressive strength was exceeded. When cracking occurred, it was modelled through an adjustment of material properties which effectively treats the cracking as a "smeared band" of cracks. The stress-strain matrix was adjusted by introducing a plane of weakness in a direction normal to the crack face, and two shear transfer coefficients for open and closed cracks, $\beta_t = 0.15$ and $\beta_c = 0.75$ were considered. The β_t represents a shear strength reduction factor for those subsequent loads which induce sliding – shear – across the crack face. If crushing occurs, the complete deterioration of the structural integrity of the material is considered and the contribution to the stiffness of an element at the integration point is ignored.

The failure criterion was defined by means of two uniaxial strengths, namely uniaxial compressive strength, f_c, and uniaxial tensile strength, f_t. A parameter variation study of results concerning the uniaxial tensile strength as a percentage of the compressive strength was performed. The value thus selected was consistent with the present conservation state. With regard to the Drucker-Prager parameters, the expressions proposed by Lourenço [15] were considered, as shown in table 2.

Table 2: Model calibrating parameters.

Parameter	Material criteria				
	Elastic	Drucker-Prager	Willam-Warncke	Nonlinear c c	Trilinear
Young's modulus, E (Pa)	1.e9			1.e9	
Poisson's ratio, v	0.2			0.2	0.2
Specific weight, w (N/m^3)	17,000			17,000	17,000
Shear transfer coefficient β_t			0.15		
Shear transfer coefficient β_c			0.75		
Compressive strength, f_c (Pa)			1.e6	1.e6	1.e6
Tensile strength, f_t (Pa)			1.e5	1.e5	
Strain at peak stress, ε				0.002	
Confinement				1	
Initial stiffness, E_0 (Pa)					1.e9
Post-peak stiffness E_f (Pa)					−1.e8
Residual strength, f_R (Pa)					1.e5
Cohesion (Pa)		1.5e5			
Internal friction		54°			
Dilatancy		10°			

The results thus obtained were consistent with the linear analysis. Maximum compressive stresses – 0.36 MPa – were reached by the basement, as expected. Maximum tensile stresses – 0.17 MPa – appeared in the vaults, top level of the core and windows, as well as, high stress concentrations were obtained in the connections among structural parts. The compressive stresses are admissible, but stresses higher than tensile strength caused cracks, as fig. 4 shows. With regard to displacements, the maximum horizontal values were reached by the vaults and the core.

5 Modal analysis

A modal analysis was performed on the aforementioned refined elastic three-dimensional finite element model, in order to obtain the dynamic properties – natural frequencies, ω_n, and modal shapes, ζ_n – and to serve as a starting point for the transient dynamic analysis. Damping was not taken into account. The sum of the effective modal masses for the considered modes is more than the 90% of the total mass – 133,208.49 kg. All these modes have an effective mass greater than 5% of the total mass. The first 4 modal shapes are provided in fig. 5 and the modal participating mass ratios for each principal direction are reported in table 3.

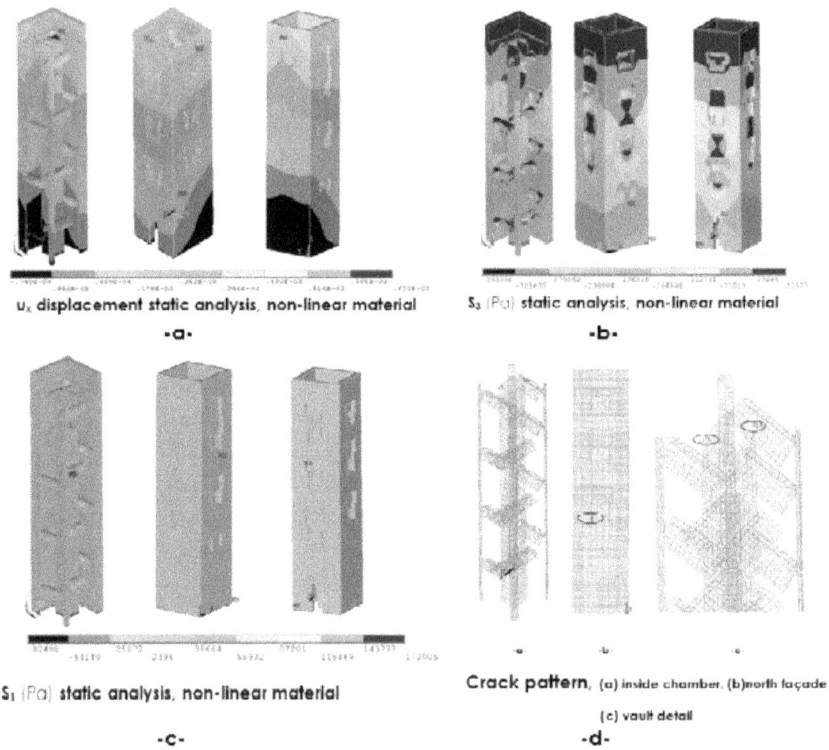

Figure 4: Static analysis with non-linear material results.

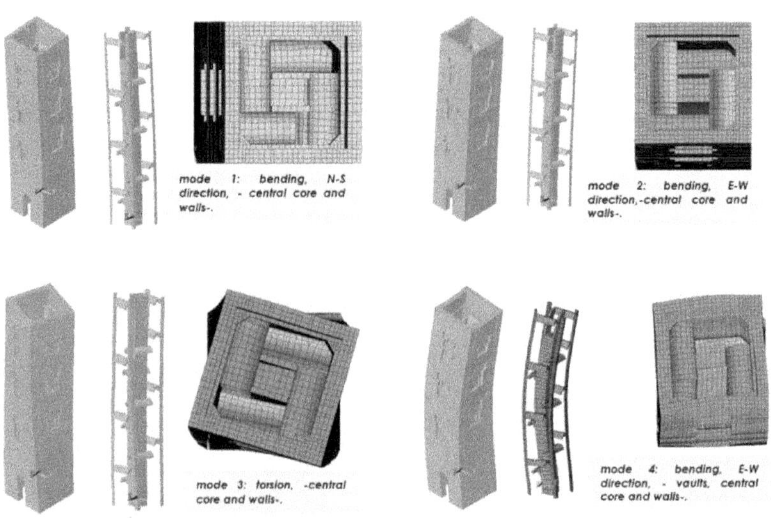

Figure 5: Modal shapes.

Table 3: Effective modal masses.

Mode	Period (s)	Frequency (Hertz)	Effective mass x direction*(kg)	Effective mass y direction*(kg)	Effective mass z direction*(kg)
1	0.49	2.05	80410.5	2.91975	2.50050
2	0.48	2.09	2.76858	81625.4	0.134283E-1
3	0.15	6.82	15.1173	1040.00	0.961566E-1
4	0.12	8.53	1.36813	28428.9	4.01811
5	0.11	8.82	28259.5	0.292487E-1	12.2214
6	0.08	11.95	41.2666	9.14584	30.9422
7	0.07	13.17	15.0521	2.22816	93431.8
8	0.07	14.14	656.086	1084.06	513.403
9	0.07	14.43	71.6396	15.5072	11306.4
10	0.06	15.09	961.888	1827.76	150.233
11	0.06	16.57	300.605	60.4455	18.2308
12	0.06	16.67	47.5473	8.36312	0.141486
13	0.06	16.79	121.263	16.0208	0.525862
14	0.06	17.27	252.967	1108.26	0.792379
15	0.06	17.67	1458.99	2553.17	0.139824
16	0.05	18.10	3228.51	276.509	1.02579
17	0.05	19.16	5.58691	367.799	0.442064E-1
18	0.05	19.53	12.9383	13.6242	0.769535
19	0.05	20.58	11.3614	1684.52	0.301165
20	0.05	21.14	1840.22	0.793902	2.52932
			ΣMeff=117,715	ΣMeff= 120,126	ΣMeff= 105,476

**x direction; parallel to east and west façades.*
y direction; parallel to north and south façades.
z direction; parallel to longitudinal axis.

From the eigenvalue analysis results, it may be concluded that the first and the second modal shapes provide the highest mass contribution, and both of them involve global bending. Those shapes are characterised by a high global stiffness and a monolithic behaviour among vaults, central core and walls. The third modal shape displays torsional response, and, when higher shapes are analysed, weak collaboration among the different structural parts is revealed, and significant out-of-plane deformations are observed.

The results thus obtained are consistent with the structural type, as it is characterised by two stiff elements – perimeter walls and core – connected by more flexible elements – barrel vaults.

6 Seismic analysis

A number of three-dimensional linear and non-linear finite element models were developed, using 3-D solid elements or 3-D beams elements. All the models assume that the masonry structure is homogeneous and in order to consider the material non-linear behaviour – including crushing and cracking – different constitutive models were used. Furthermore, beam models took into account the structural response under cyclic loading. These analyses yielded significant information on the local and global collapse mechanism predictions. Previous time-history dynamic analyses are evaluated to calibrate the non-linear response, and to take into account that crack opening may introduce numerical instabilities.

6.1 Seismic action

The Cuatrovitas tower is placed in a seismic area, and the value of the peak ground acceleration is 0.07g according to the Spanish Seismic Standard [16]. Five synthetic accelerograms were generated following the proposed method by Gasparini and Vanmarcke [17]. Those ground acceleration time-histories are compatible with the Eurocode 8 [18] horizontal and vertical design spectra. The type 2 elastic response parameters, for a ground type C, were adopted. The stationary part duration was equal to 10 s and the total duration was equal to 20 s (fig. 6). Modelling of seismic action was achieved by introducing three simultaneously acting synthetic accelerograms at the basement, which were applied at each structural principal axis. The time-dependent structural response was obtained from 5 nonlinear time-history analyses, following the component combinations proposed by EC8.

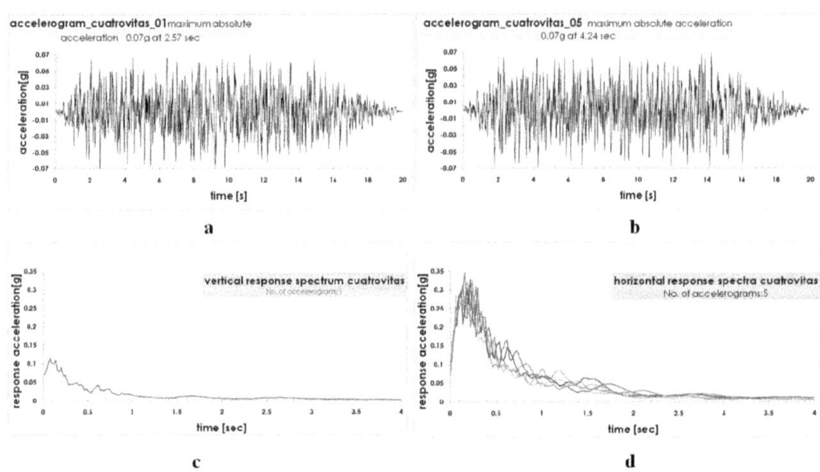

Figure 6: Seismic action.

6.2 Non-linear dynamic analysis of simplified beam models

Simplified 3-D beam models were used to predict the nonlinear inelastic response of the tower subjected to the seismic loading. Due to the geometrical properties, the component combination applied was: $0.30E_{Edx}+E_{Edy}+E_{Edz}$, where, E_{Edx}, E_{Edy} and E_{Edz} represent the action effect due to the application of the horizontal and vertical component of the design seismic action.

The direct integration of the motion equations was accomplished using the Newmark algorithm [19]. The ANSYS and SeismoStruct [20] F.E. software were used to construct the models. The BEAM4 ANSYS element and the 3-D Inelastic Frame beam-column SeismoStruct element were used. Cantilever beam models, comprised of the central solid core, were analysed in order to obtain an upper limit response. Other recent research works have considered structural parts as perfectly merged in beams models in order to obtain lower limit response [21]. The analysed tower mesh consists of 15 elements and 16 nodes. The node position is the same that the vaults position. Modelling of non-linear material behaviour is achieved by means of different constitutive models, namely, a uniaxial nonlinear constant confinement model, and a simplified uniaxial trilinear concrete model.

The former follows the constitutive relationship proposed by Mander *et al.* [22] and the cyclic rules proposed by Martinez-Rueda and Elnashai [23]. This is a cyclic stress-strain model for both confined and unconfined brittle materials, and it is able to provide a good estimation of the cyclic response of structures dominated by flexure and axial force under static and dynamic conditions. The specific model calibrating parameters used to fully describe the mechanical characteristics of this model were: compressive strength f_C, tensile strength f_t, strain at peak stress ε, confinement factor k_c, specific weight w, Young modulus E, and Poisson ratio v, see table 2. This approach allowed the assessment of the structural response under cyclic loading, taking into account inelastic strain and degradation of strength and stiffness.

The latter model assumes no resistance to tension and features a residual strength plateau. The specific model calibrating parameters used to describe the mechanical material properties were: compressive strength f_{c1}, initial stiffness E_1, post-peak stiffness E_2, residual strength f_{c2}, specific weight w, Young modulus E, and Poisson ratio v (see table 2).

A Rayleigh model for damping was assumed, that is:

$$c = a_0 m + a_1 k;\ a_0 = \zeta \frac{2\omega_i \omega_j}{\omega_i + \omega_j},\ a_1 = \zeta \frac{2}{\omega_i + \omega_j} \tag{1}$$

where c, m and k are the damping, mass and stiffness matrices, respectively. The two modes i and j were assumed to have the same damping ratio equal to 3%. That ratio is a recommended value for unreinforced masonry structures [24].The modes considered are the first one and the second one, as they contributed significantly to the response. The first natural frequency ω_1 was equal to 3.73 rad/s and the second natural frequency ω_2 was equal to 23.4 rad/s. According to eqn (1), coefficients a_0 equal to 0.19 and a_1 equal to 0.002 were obtained.

Those simplified models provided relevant information when global collapse mechanisms such us the tower over-turning were analyzed. The solid central core is safe if, for each time step the response eccentricity is less than the eccentricity of the normal force producing the over-turning. After analysing stresses and displacements it may be concluded that the tower is stable under this earthquake loading (fig. 7).

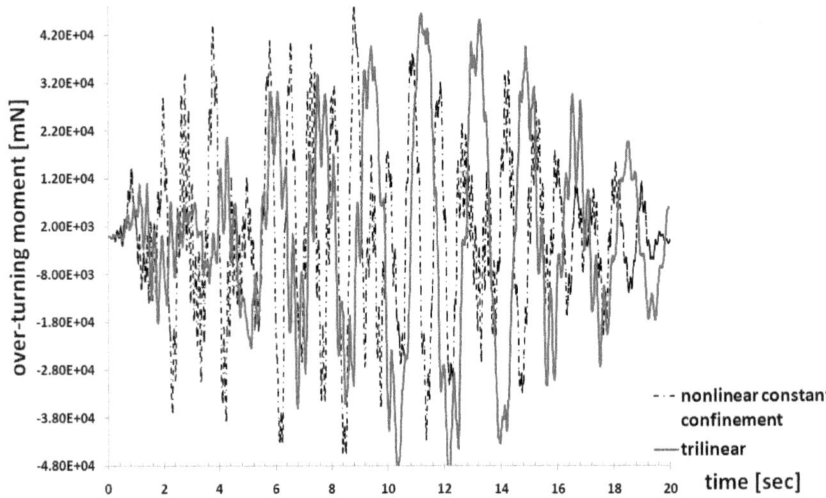

Figure 7: Over-turning assessment.

6.3 Non-linear dynamic analysis of detailed 3-D solid model

Additional dynamic time-history analysis was performed in order to predict the nonlinear inelastic response of the tower. The same 3-D FE model as in section 4.2 was used.

Table 4: Non-linear dynamic analysis results.

transient analysis results: displacements						
material	u_{xmin}(m)	u_{xmax}(m)	u_{ymin}(m)	u_{ymax}(m)	u_{zmin}(m)	u_{zmax}(m)
linear material	-0.24E-3	0. 19E-3	-0.0017	-0.52E-4	-0.0025	0.78E-4
non-linear material	0.0035	0.0028	0.01	0.01	0.007	0.005

transient analysis results: principal stresses				
material	S_{1min}(MPa)	S_{1max}(MPa)	S_{3min}(MPa)	S_{3max}(MPa)
linear material	0.1(compressive)	0.17(tensile)	0.5(compressive)	0.013(tensile)
non-linear material	0.1(compressive)	0.19(tensile)	0.59(compressive)	0.015(tensile)

A Rayleigh model for damping-according to the previous modal analysis-was assumed, and the coefficients a_0 equal to 0.69 and a_1 equal to 0.0009 were applied. Two different materials were considered, namely the linear elastic material and the Drucker-Prager Willam-Warncke non-linear material. The results obtained show that the low tensile strength exhibited by masonry, leads to a higher cracking pattern than that predicted by the static approach. The vaults, the framed small openings along the external walls, and the inside walls show signs of wear. The largest tensile stresses are reached in the connections of the barrel vaults with the inside walls and the central core. The maximum compressive stress is far below the compressive strength, and obviously, no crushing is expected. The displacement range is quite low, and no over-turning is caused. In spite of the dynamic input the structure exhibits a monolithic behaviour. The obtained results are provided in table 4 and the cracking patterns are reported in fig. 8.

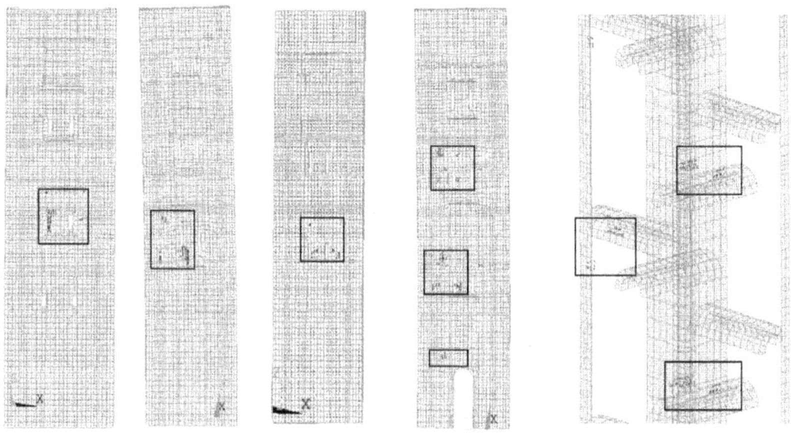

Figure 8: Cracking patterns.

7 Conclusions

The Cuatrovitas medieval tower has been analysed under seismic loading, in order to assess its structural behaviour. For this purpose, a number of three-dimensional linear and non-linear finite element models with different levels of complexity and simplifications were developed.

As a summary of the results, it may be concluded that static analyses, with linear and non-linear materials, provide relevant information regarding to interaction and stress distribution among structural parts. Furthermore, weak points which are prone to suffer damage under seismic loading may be identified – for instance, vaults, windows and connections.

As far as the seismic transient analysis is concerned, it may be stated that the Drucker-Prager perfectly plastic criterion and the Willam-Warncke failure surface, are valuable tools in order to predict local failure – cracking progression – in a detailed 3-D FE model. Those criteria yield reliable results when neither crushing nor high plastic deformation due to high compressive stresses are expected. Moreover, as the low tensile strength exhibited by the material yields in non-linear behaviour, it is important to use an effective cracking constitutive model.

The simplified models allow the determination of the upper bound response – that is, global collapse mechanisms such as tower over-turning - taking into account the response under cyclic loading. Two constitutive models, namely, a uniaxial nonlinear constant confinement model and a simplified uniaxial trilinear concrete model were applied; both of them provide significant data in good agreement.

The aforementioned methodology might be useful in assessing safety and improving seismic resistance of this and other similar cultural heritage buildings.

References

[1] Lourenço, P.B. & Roque, J.A., Simplified indexes for the seismic vulnerability of ancient masonry buildings. *Construction and Building Materials,* **20**, pp. 200–208, 2006.

[2] Binda, L., Saisi, A. & Tiraboschi, C., Investigation procedures for the diagnosis of historic masonries. *Construction and Building Materials,* **14**, pp. 199–233, 2000.

[3] Gurriarán Daza, P., *Acerca del alminar almohade de Cuatrovitas en Bollullos de la Mitación (Sevilla)*, Delegación de Cultura, Ayto. Bollullos de la Mitación. 2000.

[4] Bloom, J., *Minaret, Symbol of Islam*, Oxford Studies in Islamic Art VII, Oxford University Press, pp. 99–125, 1989.

[5] Valor Piechotta, M. *Aún más sobre Cuatrovita: Análisis de sus fuentes documentales y prospección arqueológica.* Estudios de Historia y Arqueología medieval, II, Cádiz, 1983.

[6] ANSYS, ANSYS Manual SET, ANSYS Inc., Southpoint, 275 Technology Drive, Canonsburg, PA 15317, USA, 1998.

[7] Barrios Padura, A., Estudio de las cimentaciones de edificios históricos en la provincia de Sevilla: siglo XII al XVI. Tesis Doctoral, Universidad de Sevilla, Departamento de Construcciones Arquitectónicas, 2001. (PhD Dissertation).

[8] Drucker, D.C. & Prager, W. Soil mechanics and plastic analysis or limit design. *Quarterly of Applied Mathematics*, **10**, pp. 157–16, 1952.

[9] Willam, K.J. & Warncke, E.D. Constitutive model for the triaxial behaviour of concrete. *Proceedings of the International Association for Bridge and Structural Engineering*, **19**. ISMES. Bergamo, Italy, 1975.

[10] Pallarés, F.J., Agüero, A. & Ivorra, S., A comparison of different failure criteria in a numerical seismic assessment of an industrial brickwork chimney. *Materials and Structures*, **42**, pp. 213–226, 2009.

[11] Oliveira, D.V. & Lourenço, P.B., Implementation and validation of a constitutive model for the cyclic behaviour of interface elements. *Computers and Structures*, **82**, pp. 1451–1461, 2004.

[12] Fanning, P.J. & Boothby, T.E., Three-dimensional modelling and full-scale testing of stone arch bridges. *Computers and Structures*, **79,** pp. 2645–2662, 2001.

[13] Bartoli, G., Betti, M., Spinelli, P. & Tordini, B., An “innovative” procedure for assessing the seismic capacity of historical tall buildings: The “torre Grossa” masonry tower. *Structural Analysis of Historical Constructions*, eds. P.B. Lourenço, P. Roca, C. Modena & S. Agrawal, New Delhi, 2006.

[14] Betti, M. & Vignoli, A., Assessment of seismic resistance of a basilica-type church under earthquake loading: Modelling and analysis, *Advances in Engineering Software*, **39(4)**, pp. 258–283, 2008.
[15] Lourenço, P.B., Computational strategies for masonry structures, Dissertation, Delft University of Technology, Delft, The Netherlands. 1996.
[16] Norma de Construcción Sismorresistente: Parte General y Edificación. NCSE-02, Ministerio de Fomento, (Spanish Standard). 2002.
[17] Gasparini, D.A. & Vanmarcke, E.H., *Simulated earthquake motions compatible with prescribed response spectra*. Massachusetts Institute of Technology, Cambridge, Massachusetts, 1976.
[18] Eurocode 8. Design provisions for earthquake resistance of structures. Part 1–4: General rules – strengthening and repair of buildings. ENV 1998-1–4; 2004.
[19] Newmark, N.M., A method of computation for structural dynamics. *Journal of the Engineering Mechanics Division, ASCE,* **85**, No.EM3, pp. 67–94., 1959.
[20] SeismoStruct [2007]. "SeismStruct – A computer program for static and dynamic nonlinear analysis of framed structures" [online]. Available from URL: http://www.seismosoft.com.
[21] Project Title: Improving the Seismic Resistance of Cultural Heritage Buildings, Project Contract N°: ALA/95/23/2003/077-122, Title: Benchmarking on the seismic behaviour of the Qutb Minar, Project Beneficiary: Universidade Do Minho, Portugal. Asia-wide Programme: EU-India Economic Cross Cultural Programme, 2006.
[22] Mander, J.B., Priestley, M.J.N. & Park, R., Theoretical stress-strain model for confined concrete. *Journal of Structural Engineering,* **114(8)**, pp. 1804–1826, 1988.
[23] Martinez-Rueda, J.E & Elnashai, A.S., Confined concrete model under cyclic load. *Materials and Structures*, **30(3)**, pp. 139–147, 1997.
[24] Chopra, A.K. *Dynamics of structures: Theory and Applications to Earthquake Engineering*, Prentice-Hall, 1995.

The effect of earthquake characteristics on the collapse of historical masonry buildings: case study of the mosque of Takiyya al-Sulaymaniyya

W. Jäger & T. Bakeer
Chair of Structural Design, Dresden University of Technology, Germany

Abstract

Different factors influence the collapse behaviour of masonry structures and one of the major factors is the characteristics of the earthquake itself. This effect is going to be more complicated for historical constructions, which involve different structural members with different geometries. The present study focuses on the effect of earthquake characteristics on the collapse behaviour of historical masonry structures. The effect of earthquake direction and the frequency content of the earthquake are discussed. A brief background of the selected case study and modelling process are given first and the geometry of the whole structure is created where micro modelling strategy is employed. The collapse analysis of the structure is performed under an artificial model based on the earthquake characteristics. Unidirectional earthquakes are applied to the structure from different angles in order to investigate the weakest situation. Finally, different earthquake models are generated with different frequency contents according the soil profiles and applied consequently on the structure to explore the worst situation.
Keywords: collapse analysis, historical masonry buildings, earthquake direction, frequency content, artificial accelerogram.

1 Introduction

The architecture of historical masonry structures shows a wide disparity through centuries and the masonry structural members are formed in various geometries (pillars, arches, vaults, domes and minarets). However, the variation in the geometries of such structural elements also results in different performances against earthquakes and their vulnerability to collapse. The study of the behaviour of a single structural element may give an indication about its individual vulnerability to earthquakes. However, the success of such study is associated with considering the other existing structural members that shape the entire geometry. For example, domes are intrinsically much stronger against earthquakes than other members and their possible weakness is essentially associated with the stiffness and strength of the supporting members. In the case of Hagia Sophia, the deformability of the main pillars and supporting arches are caused sometimes by the damage and the collapse [3]. For these reasons, the collapse of historical masonry structures will be demonstrated through a case study of a full masonry structure that comprises the disparities in structure elements (pillars, arches, vaults, domes and minarets). Indeed, many masonry structures exist in the world that have such a variety in elements, but those that are in earthquake hazardous regions are of more relevance to this study.

Good examples of such buildings are conceivably those built by the medieval architect of the Ottoman Empire Mimar Sinan. Furthermore, most of those buildings are constructed in regions that experience seismic activity such as Turkey, Syria, Greece, Cyprus, Ukraine and Bulgaria. Among the large number of the works by Sinan, the beautiful mosque of Takiyya al-Sulaymaniyya in Damascus, the capital city of Syria, has been chosen for the present study.

2 Mosque of Takiyya al-Sulaymaniyya

Takiyya al-Sulaymaniyya is a complex in Damascus, Syria, considered as the most important Ottoman cultural building in the city. It was built by the Sultan Süleyman I or Sulayman al-Qanuni (1520–1566) between 1554 and 1560 [7]. The entire complex was restored in the 1960s by the Directorate General of Antiquities of Syria.

The mosque is the largest and the major part of the complex, located on the southern end of the courtyard. The architecture of the mosque is similar to the prototypical forms used by Sinan, a cubic mass crowned by a vast hemispherical dome rising over pendentives, with a portico in front and twin minarets. The hall of the mosque is based on a square plane of 16 by 16 meters. The dome is suited on a square of 14.3 by 14.3 meters and rises to a height of 7.4 meters on a circular base supported by four pendentives and four large arches which stand at the corners of the square bases (figs. 1 and 2). The diameter of the hypothetical sphere which includes both the dome and the pendentives (the diagonal of the square) is about 20.1 meters.

The domes in the interior portico rest on pendentives (fig. 3), whereas the central bay over the mosque entrance is roofed by a vault higher than the domes. The exterior portico is covered by a shed roof. The pointed arches in exterior and interior porticos are braced using steel bars. This kind of bracing system is commonly used in most works of Sinan.

Figure 1: North-east (left) and south-east (right) views for the mosque.

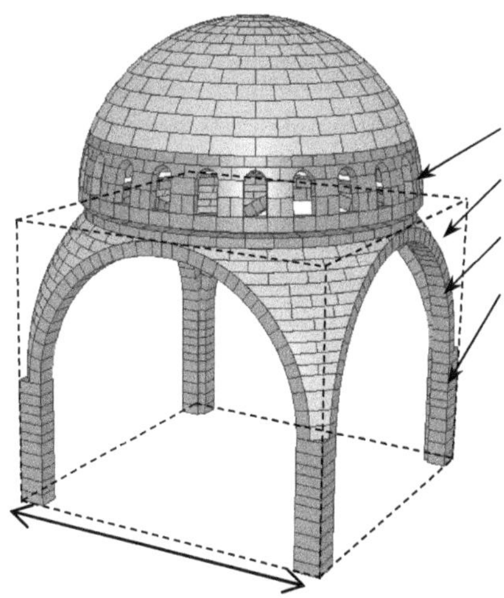

Figure 2: The dome of the mosque supported on pendentives.

The mosque of Takiyya al-Sulaymaniyya has twin high polygonal minarets. The first high storey of the minaret ends with a balcony. The second short storey is 9.4 m in height and covered by a typical conical crown, sheltered by lead.

3 Finite element modelling

The whole building is modelled in LS-DYNA software stone by stone. The stone cuts are considered in modelling. The mesh of each stone has been generated, so that, stones of simple geometries are meshed by 3×3×3 elements. Finer meshes lead to a large number of elements in the overall structure and courser meshes possibly cause numerical instabilities in contact treatments using the LS-DYNA code. The constant stress eight-nodded brick element with a single integration point has been employed with a Flangan-Belytschko stiffness form to control the hourglass effect [4].

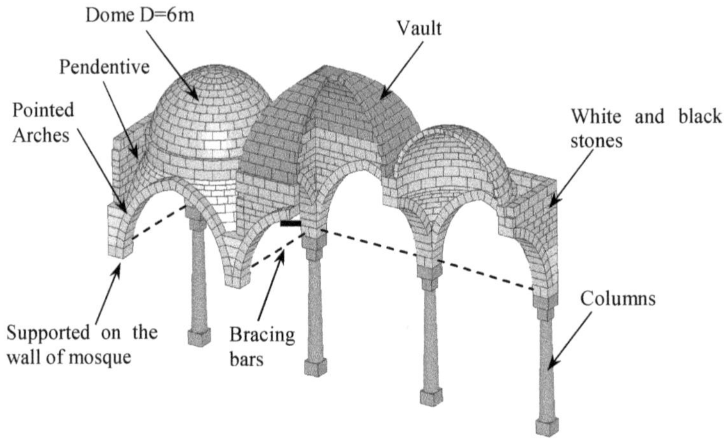

Figure 3: The interior portico in front of the mosque.

Each part of the structure is created in a separate file that contains the finite elements of this part. Later, the whole parts are assembled in LS-PrePost. The elastic material model is adopted in the following study, due to the relatively high material strength. This model guarantees a smooth running through the calculation and avoids the termination due to negative volumes, which arises with soft materials. The tiebreak contact model is employed to represent the interface between the units. The tiebreak contact in LS-DYNA allows the modelling of connections, which transmit both compressive and tensile forces with an optional failure criterion [2]. The option 6 of contact tiebreak permits damage modelling by scaling the stress components after failure is met. The following yield function is employed:

$$F(\sigma,\kappa)=\frac{\sigma^2}{f_t^2}+\frac{\tau^2}{c^2}-\Omega(\kappa) \tag{1}$$

where f_t is the tensile strength of the interface, c is the shear strength of the interface, $\Omega(\kappa)$ is the linear damage function, given by $\Omega(\kappa) = 1 - \kappa$, κ the damage scalar given by $\kappa = w/w_c$ and w, w_c the current crack width and crack width at failure, respectively. After the damage is initiated, the stress is linearly scaled down until the crack width reaches the critical distance w_c at which the interface failure is complete.

The resulting model comprises 768887 nodes and 363567 elements. Therefore, parallel processing is adopted. Each calculation is performed using parallel 40 Intel Itanium processors “SGI Altix 4700” in the centre of High Performance Computing of TU-Dresden. The calculation for the model was carried out along 7 days for 20 seconds of loading.

4 Earthquake modelling

In Syria, during the last decades, the region has experienced several destructive earthquakes and it is one of a few places worldwide where historical strong earthquake events are well documented [5, 9].

The Syrian engineering code provides a PGA distribution map for minimum 10% probability of being exceeded in a life time of 50 years, which corresponds to return periods of 475 years. This PGA distribution map is used for the design of engineering structures. However, for historical monuments a longer life time of up to 200 years must be considered.

Malkawi *et al.* [5] employed a probabilistic model that makes possible to estimate the probability of occurrence of forthcoming earthquakes in Syria, based on available information on seismicity, geo-tectonics setting and attenuation characteristics of peak ground acceleration. Several PGA distribution maps are proposed by Malkawi *et al.* [5]; the map of PGA distribution for 10% probability of being exceeding in a life time of 200 years is plotted in fig. 4. In addition to the regional seismicity, further information can be obtained from the design response spectra provided in engineering standards. The spectral analysis of engineering buildings in Syria usually follows the response spectrum provided in UBC97 [11]. This response spectrum is related to the site specific values of C_a, C_v, which are given in Table 1.

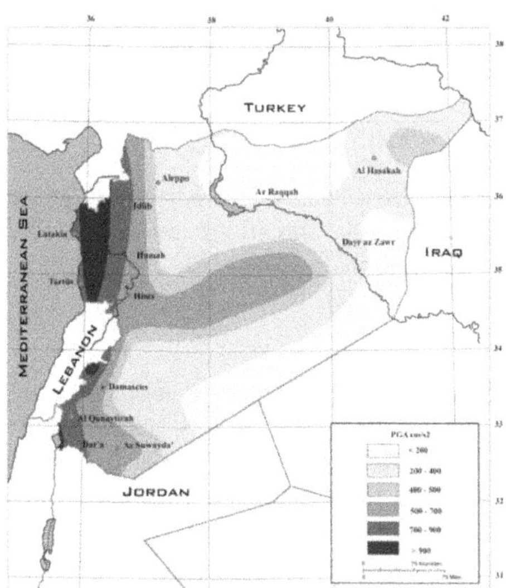

Figure 4: Maximum peak ground acceleration (cm/sec^2) with 10% probability exceeding in a life time of 200 years (return period=1898 years).

Since there are no earthquake records for strong intensities, the generation of an artificial one is needed. Many methods proposed in the literature to generate an artificial accelerogram for regions that are lacking earthquake records, namely: the sums of harmonic functions, filtering of white noise [6, 10], the spectral density model [8] or finite element modelling of the fault system [1]. The sum of harmonic function method is employed in this study to generate artificial accelerograms.

Table 1: Values of C_a, C_v according to the soil type [11].

Soil Type	Soil profile name/ Generic description	Share wave velocity	C_a	C_v
S_A	hard rock	>1500	0.32	0.32
S_B	rock	760 to 1500	0.40	0.40
S_C	very dense soil and soft rock	360 to 760	0.40	0.56
S_D	stiff soil profile	180 to 360	0.44	0.64
S_E	soft soil profile	<180	0.36	0.96

5 Effect of earthquake characteristics

The collapse analysis for a specific earthquake motion gives a very crude estimation of the behaviour under prospective earthquakes. The collapse analysis under the actions could emphasize the weak parts of the structure, but the weak state is also associated with loading conditions. In the case of earthquakes, there are high uncertainties. However, several parameters were provided in quantitative form to characterize the random motion of earthquakes that might influence the behaviour of the structure, like Peak ground acceleration and incremental velocities. The frequency content of the ground motion is also of high importance.

In the following, the collapse analysis is performed in order to understand the response for different earthquake directions, as well as to explore the effect of frequency content of the earthquake.

5.1 Direction of the earthquake

The vertical component of the earthquake has less effect on the structure, due to the safety margins against the static gravity acceleration. However the peak vertical acceleration is often assumed to be 2/3 of the peak horizontal acceleration [12]. To study the earthquake component in the horizontal plane, two principal directions can be considered. The first is associated with the structure '*The principal direction of the structure*' and indicates the weakest direction of the structure, and the other is associated with the earthquake '*the principal direction of the earthquake*' and corresponds to the direction in which the horizontal ground acceleration amplifies the maximum. The worst case occurs when the principal direction of the structure and the earthquake are identical.

Due to the uncertainties regarding the earthquake direction and in order to achieve a good assessment, the structure should be capable of equally resisting earthquake motions from all possible directions. According to some of the existing engineering standards, the structure should be assessed for "100% of the prescribed seismic forces in one direction plus 30–40% of the prescribed forces in the perpendicular direction" [12]. However, no suggestions are made on how the directions have to be determined for complex structures. In order to understand the collapse behaviour of the structure for different earthquake

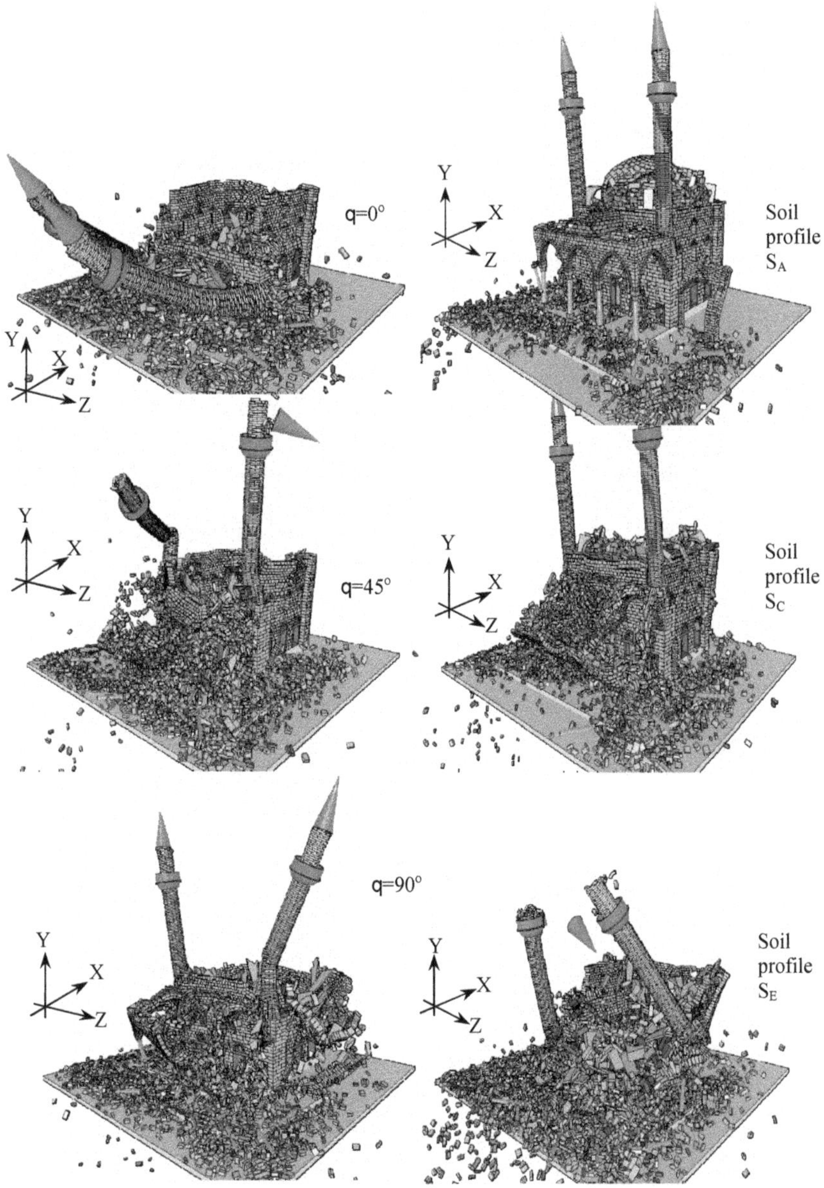

Figure 5: The collapse states of the structure for different earthquake angles at time 9 s (right), for different soil profiles at time 7 s (left).

directions, the same generated accelerogram has been applied on the structure in the horizontal plane, for different angles with respect to x axis, namely: q=0°, 45° and 90°. Fig. 5 shows the collapse states of the structure which correspond to the angles q=0°, 45° and 90° after 9 s from the initiation of the earthquake. It is quite evident that the weakest case of the structure corresponds to q=0°, when the earthquake is more destructive than for the other directions.

5.2 Frequency content of the earthquake

The earthquake actions span a broad range of frequencies. The frequency content describes how the amplitude of the ground motion is distributed among different frequencies. A good description of this relation for an earthquake can be obtained from the corresponding response spectrum.

Due to the significant influence of the frequency content of an earthquake ground motion on the structure, the latter has been subjected to different earthquakes with different frequency contents. The geological properties of the site are highly influencing the frequency content of the earthquake motion that the structure receives. Therefore, three earthquake motions were generated for different soil profiles, namely: S_A, S_C and S_E (fig. 6). The aim of considering several soil profiles in this study is to understand the influence of the site characteristics on the collapse behaviour of the structure. Furthermore, many structures in several countries have architecture similar to that of the mosque of Takiyya al-Sulaymaniyya. Fig. 5 shows the collapse states of the structure which corresponds to the soil profiles S_A, S_C and S_E after 7 s. It is quite evident that the weakest case of the structure corresponds to the soft soil profile S_E, where for this site conditions, the earthquake is more destructive.

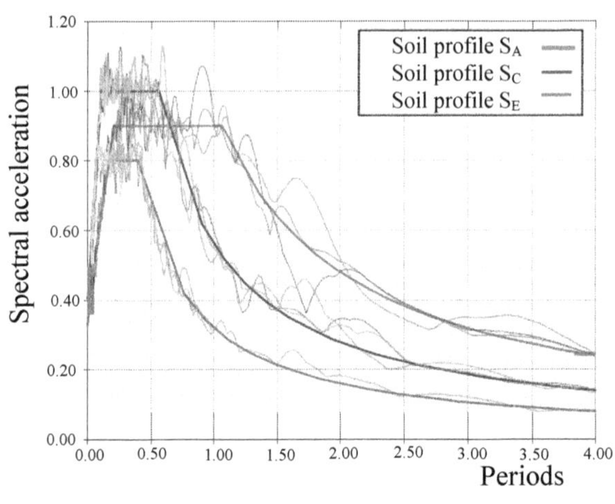

Figure 6: Response spectra for different soil types A, C and E for Z=0.4, the thick lines are the original response spectra, whereas the thin lines refer to the response spectra of the generated accelerograms.

6 Concluding remarks

Historical masonry structures which comprise different types of structural members result in different responses and therefore different collapse mechanisms. The interactions between those members are of high importance for the whole response of the structure. For the historical structure in this study, the relatively high deformations of the pendentives are the major reason for the collapse of the dome.

A collapse analysis of the structure under unidirectional earthquake actions is performed, to explore the weakest direction of the structure. However, the formulation of the mechanical behaviour is not covered in this paper, and it is very appropriate to employ the described methodology as a tool for further research in this direction. The present study shows that the earthquakes in regions of soft soils are more destructive to the structure, but the soft soils dissipate a great amount of the kinetic energy that is released by an earthquake. Furthermore, other phenomena that might occur due to the failure of the soil and liquefaction should be considered for a more detailed study.

References

[1] Aagaard, B.T., *Finite-Element Simulations of Earthquakes*, PhD Thesis, California Institute of Technology, Pasadena, California, 2000.
[2] Bala, S., Tie-Break Contacts in LS-DYNA. Livemore Software. USA 2007.
[3] Croci, G., Seismic behaviour of masonry domes and vaults of Hagia Sophia in Istanbul and St. Francis in Assisi. *Proc. of the 1st European Conference on Earthquake Engineering and Seismology*: Geneva, Switzerland 2006.
[4] Hallquist, J.O., *LS-DYNA Theory Manual*, Livemore Software Technology Corporation: California, USA 2005
[5] Malkawi, A.H., Liang, R.Y., Nusairat, J.H. & Al-Hamoud, A.S., Probabilistic seismic hazard zonation of Syria. *Natural Hazards*, **12**, pp. 139–151, 1995.
[6] Meskouris, K., *Structural Dynamics, Models, Methods, Examples*, Ernst & Sohn: Berlin, 2000.
[7] Rihawi, A., *Arabic Islamic Architecture in Syria*, Damascus, Ministry of Culture and National Heritage: Syria 1979.
[8] Rofooei, F.R., Mobarake, A. & Ahmadi, G., Generation of artificial earthquake records with a nonstationary Kanai–Tajimi model. *Engineering Structures*, **23 (7)**, pp. 827–837, 2001.
[9] Sbeinati, M.R., Darawcheh, R. & Mouty, M., The historical earthquakes of Syria: an analysis of large and moderate earthquakes from 1365 B.C. to 1900 A.D. *Annals of Geophysics*, **48 (3)**, pp. 347–435, 2005.
[10] Thiele, M., *Spectrum Compatible Accelerograms in Earthquake Engineering*. Universidad de La Serena, Chile and Technische Universität Dresden: Dresden 2002.

[11] UBC97: 1997: *Uniform Building Code, Vol. 2*. International Code Council: USA 1997.
[12] Wilson, E.L., *Three-Dimensional Static and Dynamic Analysis of Structures* – Third Edition, Computers and Structures, Inc.: Berkeley, California, USA 2002.

Numerical models to predict the creep behaviour of brickwork

A. Taliercio
Politecnico di Milano, Italy

Abstract

The numerical modelling of the time evolution of stresses and strains in brickwork under sustained loads is dealt with, within the framework of linear viscoelasticity. Finite element analyses were carried out, using three different masonry models: a simplified 2D layered model, and two 3D models (one for header bond masonry; one for Flemish bond masonry). The creep behaviour of the component materials (bricks and mortar) was defined according to experimental data available in the literature. These results were best fitted by Prony series, and the obtained creep laws were employed to carry out FE analyses of the masonry walls with different textures. Owing to the different mechanical response of the components to sustained loads, the stress and strain distribution in the wall changes in time and differs from that at the beginning of the loading process. The different behaviour of the two considered brick patterns is pointed out. The possibility of applying the simplified layered model instead of refined 3D models to predict the time evolution of stresses and strains is discussed.

Keywords: masonry, creep, linear viscoelasticity, header bond, Flemish bond.

1 Introduction

The application of sustained loads on masonry structural elements induces creep phenomena, accompanied by a redistribution of stresses and strains. Under service loads, strains usually stabilize after a given time. On the contrary, damage effects (i.e. microcracks) induced by heavy loads can coalesce and grow, bringing the structural element to failure. In the former case, anyway, the long-term stress and strain state can differ, even considerably, from the short-term one.

An overview of the mechanical (rheological) models currently employed to describe the creep behaviour of masonry and other quasi-brittle materials can be found in [1]. In the present work, only the response of brickwork under service loads will be investigated; accordingly, mortar joints and units will be assumed to have a linear viscoelastic behaviour.

The layout of this paper is as follows. First the mathematical modelling of the creep behaviour of the masonry constituents is dealt with, within the framework of linear viscoelasticity (section 2.1). The selected model is calibrated using creep test data obtained by other authors on brick and mortar samples (section 2.2). 3D finite element models of two walls characterized by different brick textures are developed and subjected to numerical creep tests; a simplified 2D layered model is also considered (section 3.1). The time evolution of the wall displacements predicted by the different models is compared with the experimental one to validate their reliability (section 3.2). The influence of the brick pattern on the time evolution of the stresses is analyzed in section 4; comparisons with the predictions given by the simplified 2D model are also added. Finally, the main findings of the work are summarized in section 5.

2 Mathematical modelling of experimental creep tests

2.1 Mathematical modelling

Within the design stress range, the components of brick masonry (units and mortar) can be individually assumed to have a linear viscoelastic behaviour. Provided that loads are applied when the structural element is sufficiently hardened, the mechanical properties of the brickwork constituents are essentially invariant in time, so that they can be assumed to be non-ageing materials.

For linear viscoelastic non-ageing materials subjected to any prescribed uniaxial stress history starting at time $t = 0$, according to Boltzmann's superposition principle the strain at any time $t > 0$ is given by (see e.g. [2])

$$\varepsilon(t) = \int_0^t J(t-\tau)\dot{\sigma}(\tau)d\tau, \tag{1}$$

where J is called creep compliance. Provided that time derivatives are intended in the distribution's sense, stress histories including jumps can also be taken into account in eqn (1).

Conversely, if a material element is subjected to a prescribed uniaxial strain history, the stress-strain law can be written as

$$\sigma(t) = \int_0^t E(t-\tau)\dot{\varepsilon}(\tau)d\tau. \tag{2}$$

E is called relaxation function, and, accordingly, is often denoted by R in the literature. The creep compliance is the 'inverse' of the relaxation function in the sense of Carson transforms [2].

Based on the so-called generalized Maxwell rheological model [3], consisting of linear springs and dashpots, the relaxation function can be expressed in the form

$$E(t) = E_\infty + \sum_{i=1}^{n} E_i e^{-t/\tau_i} \tag{3}$$

where E_∞ is the equilibrium (or delayed) modulus, E_i are called relaxation strengths and τ_i relaxation times. Note that the instantaneous modulus is $E_0 = E_\infty + \sum E_i$. The series expression in eqn (3) is usually referred to as a Prony (or Dirichlet) series.

For isotropic materials subjected to 3D strain histories, the stress-strain law can be obtained assuming that eqn (2) holds both for the hydrostatic stress ($p\boldsymbol{I}$) and the deviatoric stress ($\boldsymbol{S}$):

$$\boldsymbol{\sigma}(t) \equiv \boldsymbol{S}(t) + p(t)\boldsymbol{I} = \int_0^t 2G(t-\tau)\dot{\boldsymbol{\eta}}(\tau)d\tau + \boldsymbol{I}\int_0^t K(t-\tau)\dot{e}(\tau)d\tau, \tag{4}$$

where $e(t)$ and $\boldsymbol{\eta}(t)$) are, respectively, the volumetric and deviatoric strain at time t, K and G are, respectively, the bulk and shear relaxation moduli, and $\boldsymbol{I}$ is the identity tensor. Both relaxation moduli are assumed to be of the form (3), namely

$$K(t) = K_\infty + \sum_{i=1}^{n} K_i e^{-t/\tau_i^K}, \quad G(t) = G_\infty + \sum_{i=1}^{n} G_i e^{-t/\tau_i^G}, \tag{5}$$

where K_∞ and G_∞ are, respectively, the long-term bulk and shear moduli. In general, the relaxation times τ_i^K and τ_i^G might differ from each other; for the sake of simplicity however, $\tau_i^K = \tau_i^G \equiv \tau_i$ ($i = 1 \ldots n$) will be assumed from here onwards.

Focusing on the material shear behaviour, the deviatoric stress at any time t can be expressed as

$$\boldsymbol{S}(t) = \int_0^t 2\left(G_\infty + \sum_{i=1}^{n} G_i e^{(\tau-t)/\tau_i}\right)\dot{\boldsymbol{\eta}}(\tau)d\tau = 2G_0\left(\boldsymbol{\eta}(t) - \sum_{i=1}^{n} g_i \boldsymbol{\eta}_i(t)\right), \tag{6}$$

where G_0 is the instantaneous shear modulus, g_i are dimensionless relaxation shear coefficients and

$$\boldsymbol{\eta}_i(t) = \int_0^t \left(1 - e^{(\tau-t)/\tau_i}\right)\dot{\boldsymbol{\eta}}(\tau)d\tau = \frac{1}{\tau_i}\int_0^t e^{-s/\tau_i}\boldsymbol{\eta}(t-s)ds. \tag{7}$$

The summation in eqn (6) is the deviatoric "creep" strain, $\boldsymbol{\eta}^{cr}(t)$, that is, the difference between the total strain and the instantaneous elastic strain, $\boldsymbol{S}(t)/2G_0$.

Similar remarks apply to the viscoelastic volumetric behaviour of the material.

The creep model described above is implemented in the commercial finite element code Abaqus, which was used in the applications presented in sections 3 and 4. Details about the numerical integration of eqn (6) can be found in [4].

2.2 Experimental results

To illustrate the creep behaviour of masonry and its constituents, reference will be made to the results carried out by Brooks [5] and reported in [6] together with a theoretical prediction of the creep curves obtained from tests on masonry walls.

Calcium silicate bricks were subjected to a sustained stress of 3 MPa for 300 days, whereas Portland cement and lime sand mortar specimens were tested for a similar period under a stress of 3.4 MPa. The results obtained are reported in fig. 1, together with a best fit of the experimental points by Prony series, eqn (3), of one or two terms. The values of the parameters defining the theoretical creep curves are listed in table 1; the volumetric creep behaviour of bricks and mortar is assumed to be defined by the same non-dimensional coefficients g_i that define their deviatoric creep behaviour (see eqn (6)).

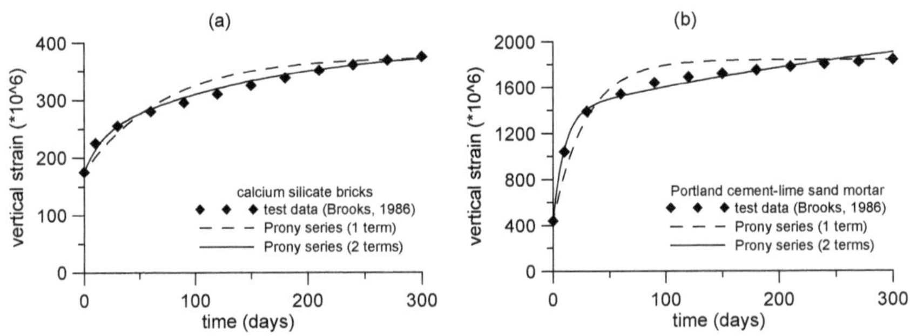

Figure 1: Creep tests on (a) calcium silicate bricks and (b) Portland cement-lime sand mortar: experimental tests [5] vs. theoretical modelling by Prony series.

Table 1: Parameters defining the creep behaviour of bricks and mortar (eq. (6)) employed in the numerical simulations.

		brick	mortar
E_0 [MPa]		17100	7700
G_0 [MPa]		7435	3208
$n = 1$	g_1	0.5327	0.7602
	τ_1 [days]	33.8	7.1
$n = 2$	g_1	0.285	0.690
	τ_1 [days]	11.1	3.1
	g_2	0.275	0.125
	τ_2 [days]	100.1	243.1

It can be noted that using Prony series of two terms, rather than one term, definitely improves the match with the experimental points. In particular, a best fit with two terms captures a primary creep phase of a duration longer than the creep test itself, conforming with the experimental data.

A single-leaf wall made of the same constituents was also subjected to a creep test of a duration of 300 days under a vertical stress of 3 MPa; the experimental results reported in [6] are shown in fig. 2. These results will be used in section 3.2 to validate the proposed finite element models.

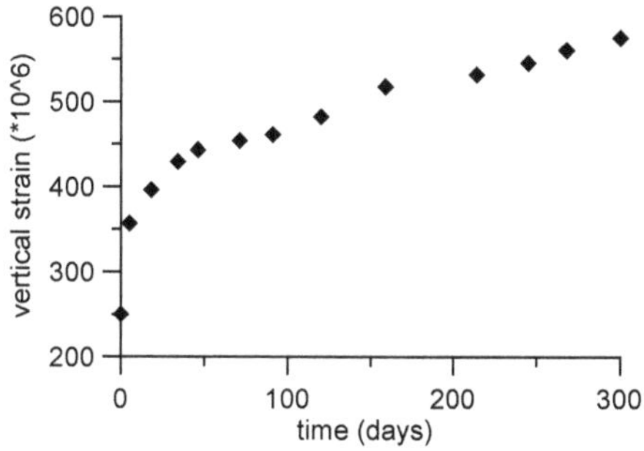

Figure 2: Creep tests on a single-leaf wall made of calcium silicate bricks and Portland cement-lime sand mortar [6].

3 Numerical simulation of creep tests on masonry walls

3.1 Finite element models of brick masonry walls

The creep behaviour of brickwork was numerically investigated using three different finite element models: a simplified two-dimensional (layered) model (fig. 3(a)) and two more detailed three-dimensional models reproducing a header bond brickwork (fig. 3(b)) and a Flemish bond brickwork (fig. 3(c)). The layered model consists of thirteen 65 mm-thick brick courses, alternating with 12 mm-thick mortar bed joints, thus matching the geometry of the brickwork subjected to the creep test shown in fig. 2 [6]. The 3D models consist of 20 brick courses; the size of the units is 55×120×250 mm^3; bed and head mortar joints have a thickness of 10 mm. The mortar-to-brick thickness ratio is 0.18 both in the 2D and in the 3D models. 4-node and 8-node isoparametric elements, with a bilinear discretization of the displacement field, were employed in the 2D and the 3D analyses, respectively.

In the applications, the self-weight of the materials is disregarded, as it was numerically found to have a negligible effect on the results. The models are subjected to a uniform vertical stress of 3 MPa at the top side; the lower base is fully restrained.

Simulations of creep tests of a duration of 300 days using creep laws described by Prony series of one or two terms were carried out. The material parameters employed in the numerical applications are those listed in table 1.

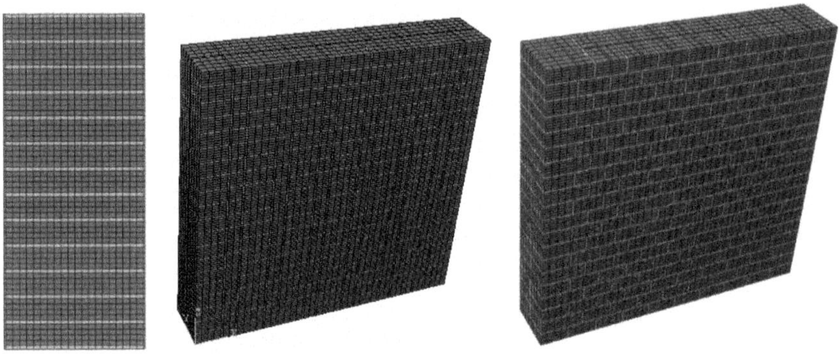

(a) (b) (c)

Figure 3: Finite elements employed in the numerical applications: (a) simplified 2D layered model; (b) 3D Flemish bond brickwork model; (c) 3D header bond brickwork model.

3.2 Creep curves

The results of the simulation of the creep test shown in fig. 2 using the 2D layered model of fig. 3(a) are shown in fig. 4, where the vertical displacement of the central node of the top side is plotted versus time. Again, using Prony series with two terms to describe the creep behaviour of bricks and mortar definitely improves the fit with the experimental data also when the global response of brickwork has to be predicted.

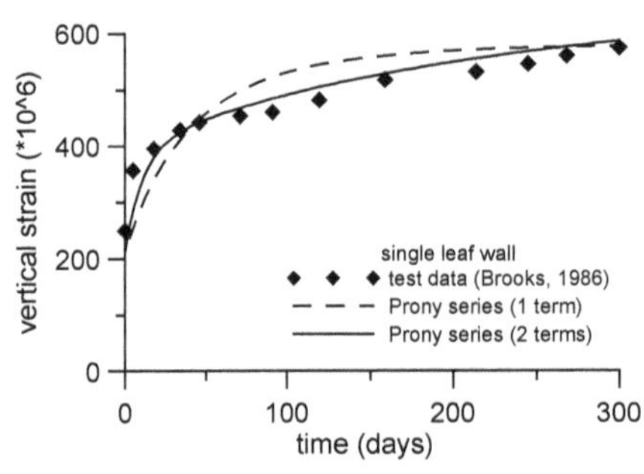

Figure 4: Creep tests on a single-leaf wall: experimental results [6] vs. predictions of the 2D FE model.

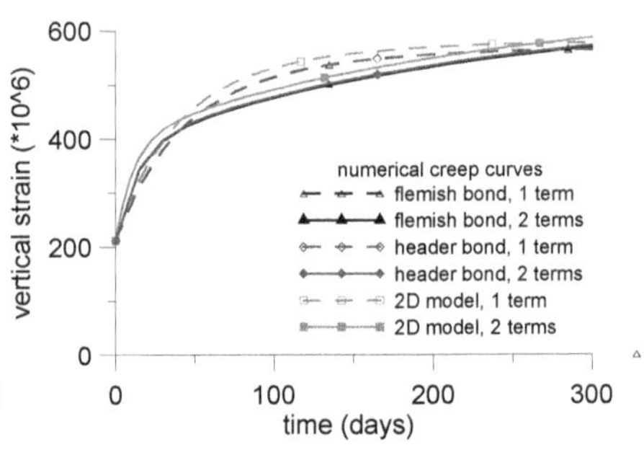

Figure 5: Simulation of creep tests on a single-leaf wall with different numerical models.

The same creep test is numerically simulated using the 3D models shown in figs. 3(b), (c), both using one term and two terms in the Prony series for the creep laws of the constituents. The results obtained are shown in fig. 5. Apparently, the difference between the Flemish bond model and the header bond model is negligible when the same number of terms in the creep law is used. The simplified 2D model and the 3D models predict the same time evolution, the 2D model being slightly more deformable than the 3D ones.

4 Stress evolution

It is instructive to follow the time evolution of the stress components in typical regions of a masonry wall, according to the different numerical models employed in the present work (see fig. 3). Only creep laws consisting of Prony series with two terms are considered.

Fig. 6 shows the distribution of the vertical stress in the 3D models at the beginning of the application of the load. The stress fluctuation induced by the brickwork heterogeneity is apparent: this fluctuation cannot be captured by the simplified layered model, which obviously predicts a uniform vertical stress of 3 MPa in both layers throughout the entire load history. The vertical stress tends to flow across the bricks bypassing the softer head joints.

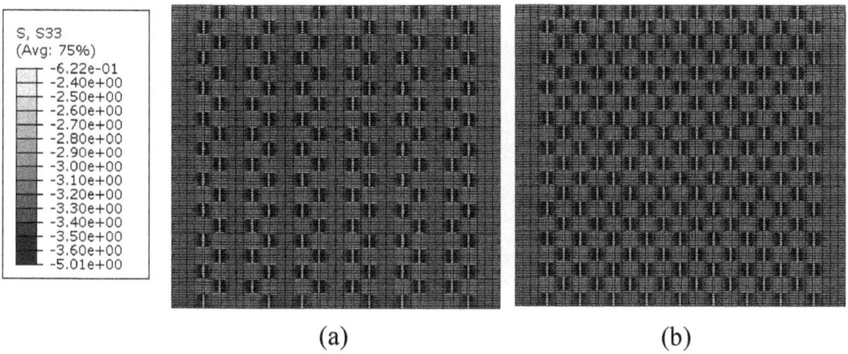

Figure 6: Vertical stress in (a) Flemish bond and (b) header bond brickwork under a uniform vertical load of 3 MPa at $t = 0$.

Fig. 7 shows the pattern of the transversal stress (i.e., the normal stress component parallel to the mortar beds) over the 3D models at the time of application of the vertical load. As is well known, owing to Poisson's effect bed joints are compressed and bricks are mostly in tension parallel to the mortar beds. The highest tensile stresses are encountered close to the head joints, which are mainly subjected to transverse tension too; these stresses might be responsible for the failure of the brick/mortar interface along these joints.

Plots of specific stress and strain components versus time, computed at different points of the models, are shown in figs. 9 to 11. All the plots refer to points in the brickwork sufficiently far from the model boundaries. As far as the layered model is concerned, points located at the mid-section of the model in the brick and in the mortar layers will be considered. The points where stresses and strains are computed in the 3D models are shown in fig. 8. These include points located in stretchers (s_1, s_2), headers (h, h_1, h_2), bed joints (bj, bj_1, bj_2) and head joints (hj). Points near the outer face of the wall (ext), as well as near the wall mid-surface (int), are considered.

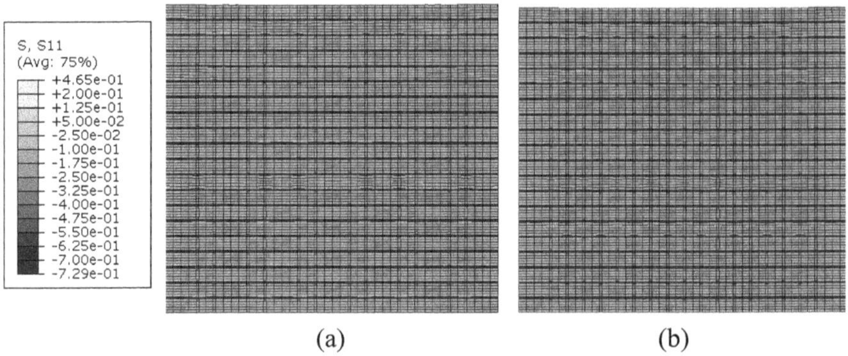

Figure 7: Transversal stress in (a) Flemish bond and (b) header bond brickwork under a uniform vertical load of 3 MPa at $t = 0$.

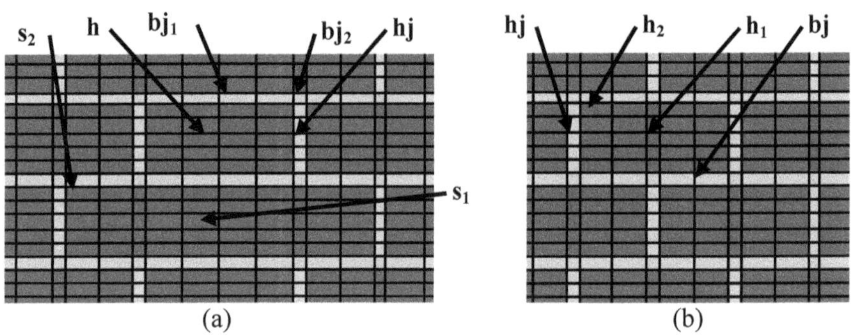

Figure 8: Points where stresses and strains are computed in (a) Flemish bond brickwork and (b) header bond brickwork.

4.1 Flemish bond brickwork

The time evolution of the vertical (axial) stress in selected points of Flemish bond brickwork is shown in figs. 9(a) and (b); fig. 9(a) refers to the points located in the bricks, whereas fig. 9(b) to points located in the bed and head joints. Note that the time evolution of the stress at most points is not monotonic; the stress attains a maximum (or a minimum) nearly 30 days after loading, and takes values close to the long-term ones at nearly 100 days after loading.

The 2D layered model underestimates the axial stress in the bricks (fig. 9(a)), except for the innermost region of the headers (curve h_1,int). The highest axial stresses are attained at the shortest sides of the stretchers, especially in the interior of the wall (curve s_2,int). The difference between the short-term predictions of the 2D model and the 3D model ranges between 2–3% in the headers (points h_1,int and h_1,ext) and 9–11% at the shortest sides of the stretchers (points s_2,int and s_2,ext). These differences increase with time; the highest discrepancy, 300 days after loading, is attained at the vertices of the stretchers (point s_2,int), and is of the order of 20% of the stress predicted by the 2D model.

In contrast, the layered model overestimates the axial stress in most regions of the mortar joints (fig. 9(b)), except for the bed joints (curve bj,ext). In particular, the axial stress predicted by the 2D model is nearly twice that computed at the head joints in the short term (points hj,int and hj,ext), and the long-term prediction is only 27% of the applied stress.

The main differences between the predictions given by the 2D model and the 3D model of Flemish bond brickwork are observed in terms of transversal stresses, which are plotted in fig. 10 versus time. Fig. 10(a) refers to points located in bricks; fig. 10(b), to points located in mortar. Contrary to the axial stress, the transversal stress predicted by the layered model varies in time, and ranges between 0.02 and 0.06 MPa in the brick layer and between –0.1 and –0.32 MPa in the mortar layer during the numerical creep test.

The 2D model underestimates the tensile transversal stress in most regions of the stretchers and overestimates the stress in the headers (fig. 10(a)). The highest computed stresses in the stretchers (curve s_2,int) vary between 0.13 MPa at t=0

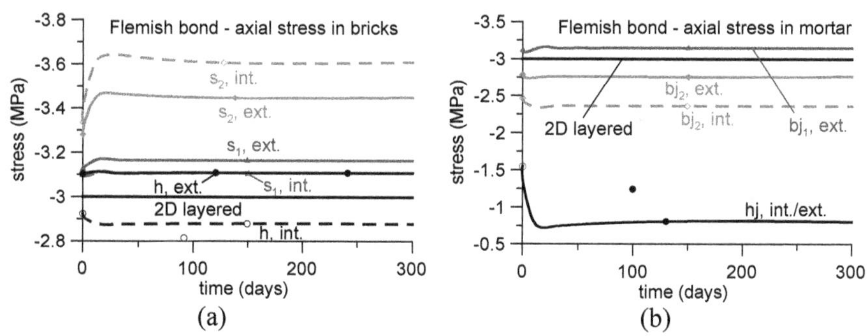

Figure 9: Time evolution of the axial stress in Flemish bond brickwork: (a) bricks; (b) mortar.

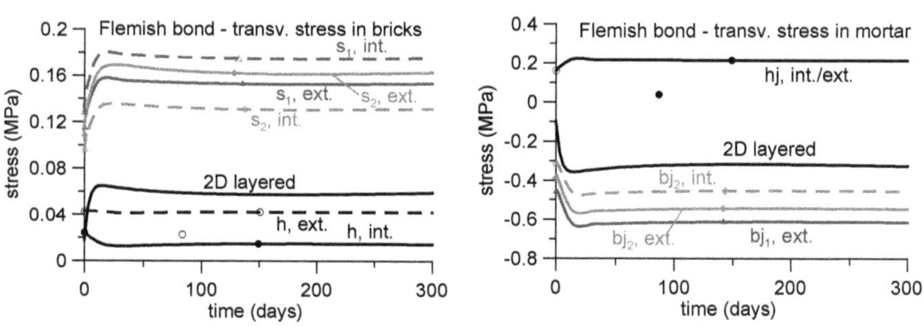

Figure 10: Time evolution of the transversal stress in Flemish bond brickwork: (a) bricks; (b) mortar.

and 0.18 MPa at t = 300 days, attaining non-negligible values contrary to the predictions of the 2D model.

The compressive transversal stress in the mortar joints is underestimated by the 2D model; the highest compressions are found in the bed joints (curve bj_1,int in Fig. 10(b)) and vary in time from –0.46 to –0.61 MPa. As already remarked, the transversal stress computed at the head joints is tensile (curve hj,int/ext in fig. 10(b)), and might cause debonding at the interface between bricks and head joints.

4.2 Header bond brickwork

Fig. 11(a) shows the time evolution of the axial stress at the points of the header bond brickwork indicated in fig. 8(b). The time evolution of the stress at most points is not monotonic, similarly to Flemish bond brickwork. The 2D layered model predicts axial stresses very close to those computed in the bed joints of the 3D model (curve bj in fig. 11(a)), and intermediate between the stresses computed at the centre (curve h_1) and at the vertices (curve h_2) of the headers. At the centre, the axial stress is lower than the applied stress by 3% in the short term, and by 5% in the long term. At the vertices, the axial stress exceeds by 9% the applied stress at the beginning of the creep test; it further increases in the long term, exceeding by 15% the applied stress. The layered model definitely overestimates the stress at the head joints (curve hj); that varies between half of the applied stress at t = 0 and 27% of the applied stress in the long term.

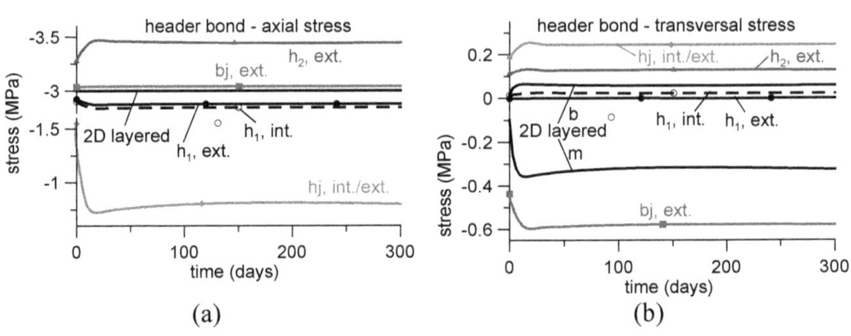

Figure 11: Stress evolution in header bond brickwork: (a) vertical stress and (b) transversal stress.

The time evolution of the transversal stress in selected points of the header bond brickwork model is shown in fig. 11(b). The stress predicted by the 2D model in the bricks falls between those computed at the centre (curves h_1,int, h_1,ext) and at the vertices (curve h_2) of the headers. Similarly to the case of Flemish bond brickwork, the 2D model definitely underestimates the compressive stress in the bed joints (curve bj): the stress predicted by the layered model is nearly half that computed by the 3D model in the long term. Again, the transversal stress in the head joints is tensile (curve hj,int/ext in fig. 11(b)), of the same order of magnitude as that computed in Flemish bond brickwork.

Finally, note that, owing to its simpler pattern, in header bond brickwork stresses are more uniform across the wall thickness than in Flemish bond brickwork (compare curves h_1,int and h_1,ext in figs. 11(a) and 11(b)).

5 Concluding remarks

The purpose of the study presented in this paper was twofold: on the one hand, assessing whether the creep behaviour of brickwork could be predicted using a simplified layered two-dimensional model rather than sophisticated 3D models; on the other hand, analysing the influence of the brick pattern on the distribution of the stresses and their redistribution in time under a constant load.

According to the comparisons discussed section 3, it can be stated that (a) the layered model predicts the experimental global creep strain of a single-leaf wall with fair accuracy, provided that two terms are employed in the Prony series that describe the creep behaviour of the individual components (see fig. 4); (b) the layered model and the three-dimensional models of two different brickworks basically predict the same axial creep strains (see fig. 5). This indicates that neither the brick pattern, nor the head joints, affect the global deformation of the wall significantly.

In contrast, the differences in vertical stresses and, even more, horizontal stresses predicted by the 2D model and the 3D models are very pronounced. This is basically due to the fluctuation of stresses (and strains) from one component to another in the same layer because of the presence of the head joints that cannot be captured by the layered model. Using the simplified 2D model might give acceptable predictions of the axial stress in the headers of the analyzed brickworks and in the bed joints of header bond brickwork, but the axial stress in the stretchers and in the bed joints of Flemish bond brickwork is heavily underestimated. The transversal stresses predicted by the 2D model are mostly unreliable. Debonding is likely to occur at the interface between bricks and head joints according to the high tensile stresses computed. To predict the time evolution of the stresses across any brickwork with more accuracy, the inability of the head joints to transfer transversal stresses should be taken into account.

Acknowledgements

The author is grateful to MM Gabriele Banfi and Pietro Gagliardi for their assistance in the finite element modelling and data processing.

References

[1] Taliercio, A., An overview of masonry creep. *Structural Studies, Repairs and Maintenance of Heritage Architecture XI*, ed. C.A. Brebbia, WIT Press: Southampton, pp. 197–208, 2009.

[2] Park, S.W. & Schapery, S.A., Methods of interconversion between linear viscoelastic material functions. Part I-a numerical method based on Prony

series. *International Journal of Solids and Structures*, **36(11)**, pp. 1653–1675, 1999.
[3] Kaliske, M. & Rothert, H., Formulation and implementation of three-dimensional viscoelasticity at small and finite strains. *Computational Mechanics*, **19(3)**, pp. 228–239, 1997.
[4] ABAQUS Theory Manual, Version 6.8-1.
[5] Brooks, J.J., Composite models for predicting elastic and long-term movements in brickwork walls. *Proceedings of the British Masonry Society*, **1**, pp. 20–23, 1986.
[6] Brooks, J.J., Composite modelling of masonry deformation. *Materials and Structures RILEM*, **23**, pp. 241–251, 1990.

Author Index

www.ingramcontent.com/pod-product-compliance
Ingram Content Group UK Ltd.
Pitfield, Milton Keynes, MK11 3LW, UK
UKHW020820290926
15470UKWH00003B/1265

9 781845 648398